Research and Perspectives in Endocrine Interactions

AF182842

For further volumes:
http://www.springer.com/series/5241

Jean-Pierre Bourguignon • Bernard Jégou •
Bernard Kerdelhué • Jorma Toppari •
Yves Christen
Editors

Multi-System Endocrine Disruption

Springer

Editors
Prof. Dr. Jean-Pierre Bourguignon
University of Liège
Developmental
Neuroendocrinology Unit, G
Place du XX Aout 7
4000 Liège
Belgium
jpbourguignon@ulg.ac.be

Prof. Dr. Bernard Jégou
GERHM Inserm U625
Campus de Beaulieu
Avenue du Général Leclerc
35042 Rennes
France
bernard.jegou@rennes.inserm.fr

PD Dr. Bernard Kerdelhué
Centre Universitaire des Saints-Pères
Inserm U648, UFR Biomédicale
75006 Paris Cedex 06
France
bernard.kerdelhue@biomedicale.univ-paris5.fr

Prof. Dr. Jorma Toppari
University of Turku
Departments of Physiology and Paediatrics
Kiinamyllynkatu 10
20520 Turku
Finland
jorma.toppari@utu.fi

Dr. Yves Christen
Fondation IPSEN pour la
Recherche Therapeutique
65 quai George Gorse
92650 Boulogne Billancourt Cedex
France
yves.christen@ipsen.com

ISSN 1861-2253
ISBN 978-3-642-27103-8 ISBN 978-3-642-22775-2 (eBook)
DOI 10.1007/978-3-642-22775-2
Springer Heidelberg Dordrecht London New York

© Springer-Verlag Berlin Heidelberg 2011
Softcover reprint of the hardcover 1st edition 2011
This work is subject to copyright. All rights are reserved, whether the whole or part of the material is concerned, specifically the rights of translation, reprinting, reuse of illustrations, recitation, broadcasting, reproduction on microfilm or in any other way, and storage in data banks. Duplication of this publication or parts thereof is permitted only under the provisions of the German Copyright Law of September 9, 1965, in its current version, and permission for use must always be obtained from Springer. Violations are liable to prosecution under the German Copyright Law.
The use of general descriptive names, registered names, trademarks, etc. in this publication does not imply, even in the absence of a specific statement, that such names are exempt from the relevant protective laws and regulations and therefore free for general use.

Printed on acid-free paper

Springer is part of Springer Science+Business Media (www.springer.com)

Preface

In the field of endocrine disruption, the reproductive system has been a priority concern of scientists and environment/public health agencies for several decades, based on observations of fertility impairment in wildlife as well as in humans. In women, intrauterine exposure to diethylstilbestrol (DES) raised the issue of carcinogenicity as early as the 1970s. Although this synthetic compound has been banned for many years, there are still unclear matters of concern such as transgenerational epigenetic effects that provide rationale for monitoring the third generation descending from exposed pregnant women. Another lesson drawn from decades of research is the widened scope of the endpoints of repro-endocrine disruption. In the male, beyond germ cell line differentiation and semen quality, other targets involve testicular descent, epididymis, prostate and external genitalia. The list of potentially disrupting chemicals has also increased dramatically: not only are pesticides and insecticides – including chlordecone and polychlorobiphenyls (PCBs) – involved but also new classes of compounds are emerging, such as analgesics. In addition, chemicals linked to plastic manufacture, like phthalates, have been studied, resulting in convincing evidence of a causal role in the rodent model of testicular dysgenesis syndrome and increasing evidence of association with the occurrence of genital malformations in human males. Although phthalates are banned from some toys and cosmetics, they are still ubiquitous. Among pending questions, the evidence linking early exposure to phthalates and oligospermia/testicular cancer in human males deserves further study. Because phthalates are present in parenteral nutrition or perfusion material, the consequences of prolonged exposure in the critical sensitive period of neonatal life need to be addressed.

More recently, the hypothalamus and the brain have been shown to be possible targets of endocrine disrupting chemicals (EDCs), accounting for neuro-endocrine disruption. In the hypothalamus, the gonadotrophin releasing hormone neurons and afferent neurono-glial system are sensitive to EDCs, leading to disorders of sexual differentiation and maturation. Aromatase and kisspeptin-neurokinin B are likely key mediators of EDC effects. Other endpoints in the CNS possibly include neurogenesis, migration and synaptogenesis in brain cortex and hippocampus. Here, the thyroid hormone system is pivotal, due to its physiological role in early

CNS development. Although they have been banned, PCBs are still among the EDCs involved, due to their persistence in the environment. A still open question is whether stresses other than chemicals (e.g., psychosocial, nutritional, etc.) interact with the sensitivity to deleterious effects of EDCs. For example, does iodine deficiency or exposure to nitrates sensitize the brain to PCB effects?

Another emerging area of endocrine disruption is the central and peripheral control of energy balance. Among other findings, the occurrence of obesity and insulin resistance in adulthood after neonatal exposure to the potent synthetic estrogen DES has opened a new field, possibly substantiating a role for EDCs, including Bisphenol A (BPA), in the epidemics of obesity and type 2 diabetes. BPA accounts for particular epidemiologic and scientific challenges due to its ubiquity and non-linear dose-response curve. The metabo-endocrine endpoints involve adipocytes, pancreatic Beta cells and the intestinal epithelium, where cell proliferation as well as differentiation could be affected by EDCs.

These three areas – repro-endocrine, neuro-endocrine and metabo-endocrine disruption – share common features: ontogenetic disturbances result from particular fetal sensitivity to endocrine disruption with sexually dimorphic responses. Epigenetic mechanisms are likely pivotal.

It has been our privilege, thanks to the Fondation Ipsen, to convene in Paris (May 9, 2011) experts to exchange findings and opinions in the different areas of endocrine disruption summarized above. Questions raised about one system may find answers based on findings in another system, justifying a multi-system perspective in endocrine disruption that appears to be a whole-body burden and a challenge for the whole scientific community, including epidemiologists, toxicologists, geneticists and endocrinologists among others.

The editors wish to express their gratitude to Mrs Astrid de Gérard for the organization of the meeting and Mrs Mary Lynn Gage for her editorial assistance.

Jean-Pierre Bourguignon

Contents

Contributors

Alonso-Magdalena, Paloma Instituto de Bioingeniería and CIBERDEM, Universidad Miguel Hernández de Elche, Elche, Spain, palonso@umh.es

Audebert, Marc UMR 1089 Xénobiotiques, INRA, Toxalim, Toulouse, France

Auharek, Sarah Centre for Reproductive Health, The Queen's Medical Research Institute, 47 Little France Crescent, Edinburgh, UK; Laboratory of Cellular Biology, Department of Morphology, Institute of Biological Sciences, Federal University of Minas Gerais, Belo Horizonte, MG, Brazil

Bourguignon, Jean-Pierre Developmental Neuroendocrinology Unit, GIGA-Neurosciences, University of Liège, Liège, Belgium, jpbourguignon@ulg.ac.be

Braniste, Viorica UMR 1054 Neuro-Gastroenteorolgy & Nutrition, Toulouse, France, viorica.braniste@toulouse.inra.fr

Brion, François Unité d'Ecotoxicologie, INERIS, Verneuil en Halatte, France, francois.brion@ineris.fr

Chung, Bon-chu Institute of Molecular Biology, Academia Sinica, Taipei, Taiwan

Drake, Amanda J Centre for Cardiovascular Science, The Queen's Medical Research Institute, Edinburgh, UK

Renato, de Franca Luiz Centre for Reproductive Health, The Queen's Medical Research Institute, Edinburgh, UK; Laboratory of Cellular Biology, Department of Morphology, Institute of Biological Sciences, Federal University of Minas Gerais, Belo Horizonte, MG, Brazil

Grandjean, Philippe Department of Environmental Medicine, University of Southern Denmark, Odense, Denmark; Department of Environmental Health, Harvard School of Public Health, Boston, MA, USA, pgrandjean@health.sdu.dk

Gore, Andrea C. The Institute for Neuroscience, Division of Pharmacology & Toxicology, College of Pharmacy, Institute for Cellular and Molecular Biology, The University of Texas at Austin, Austin, TX, USA, andrea.gore@mail.utexas.edu

Houdeau, Eric UMR 1054 Neuro-Gastroenteorolgy & Nutrition, Toulouse, France, eric.houdeau@toulouse.inra.fr

Juenger, Thomas Section of Integrative Biology, Institute for Cellular and Molecular Biology, The University of Texas at Austin, Austin, TX, USA

Kah, Olivier Neurogenesis and Estrogens, UMR CNRS 6026, Case 1302, The University of Rennes 1, Rennes cedex, France, Olivier.kah@univ-rennes1.fr

Yann, Le Page Neurogenesis and Estrogens, UMR CNRS 6026, IFR 140, Rennes cedex, France, ylepage@univ-rennes1.fr

Nadal, Angel Instituto de Bioingeniería and CIBERDEM, Universidad Miguel Hernández de Elche, Elche, Spain, nadal@umh.es

Naveau, Elise Developmental Neuroendocrinology Unit, GIGA-Neurosciences, University of Liège, Liège, Belgium, enaveau@ulg.ac.be

Newbold, Retha R. Department of Health and Human Services (DHHS), National Institute of Environmental Health Sciences (NIEHS), National Institutes of Health (NIH), Research Triangle Park, NC, USA, newbold1@niehs.nih.gov

Parent, Anne-Simone Developmental Neuroendocrinology Unit, GIGA-Neurosciences, University of Liège, Liège, Belgium, asparent@ulg.ac.be

Scott, Hayley M Centre for Reproductive Health, The Queen's Medical Research Institute, Edinburgh, UK

Sharpe, Richard M MRC/University of Edinburgh Centre for Reproductive Health, The Queen's Medical Research Institute, 47 Little France Crescent, Edinburgh EH16 4TJ, UK, r.sharpe@ed.ac.uk

Skakkebaek, Niels E. University Department of Growth and Reproduction, Rigshospitalet, Copenhagen, Denmark, nes@rh.dk

Steinberg, Rebecca M. The Institute for Neuroscience, The University of Texas at Austin, Austin, TX, USA

Tena-Sempere, Manuel Department of Cell Biology, Physiology and Immunology, University of Córdoba; CIBERobn Fisiopatología de la Obesidad y Nutrición; and Instituto Maimonides de Investigaciones Biomédicas de Córdoba (IMIBIC) and Physiology Section, Córdoba, Spain, fi1tesem@uco.es

Tong, Sok-Keng Institute of Molecular Biology, Academia Sinica, Taipei, Taiwan

Toppari, Jorma Departments of Physiology and Paediatrics, University of Turku, Kiinamyllynkatu 10, Turku, Finland, jorma.toppari@utu.fi

Sander, van Driesche den Centre for Reproductive Health, The Queen's Medical Research Institute, 47 Little France Crescent, Edinburgh, UK

Vosges, Mélanie Unité d'Ecotoxicologie, INERIS, Verneuil en Halatte, France, fabienne.carette@ineris.fr

Walker, Deena M. The Institute for Neuroscience, The University of Texas at Austin, Austin, TX, USA

Zalko, Daniel UMR 1089 Xénobiotiques, INRA, Toxalim, Toulouse, France

Zoeller, Thomas R. Biology Department and Molecular and Cellular Biology Program, University of Massachusetts Amherst, Amherst, MA, USA, tzoeller@bio.umass.edu

Neuroendocrine Effects of Developmental PCB Exposure, with Particular Reference to Hypothalamic Gene Expression

Rebecca M. Steinberg, Deena M. Walker, Thomas Juenger, and Andrea C. Gore

Abstract The production and commercial use of novel man-made chemicals over the past century have introduced a variety of industrial compounds into the environment, including polychlorinated biphenyls (PCBs). PCBs alter a multitude of physiological processes in humans and wildlife, and our laboratory and others have been studying the endocrine-disrupting mechanisms by which PCBs specifically affect the hypothalamic control of reproduction. This article will review the literature on PCB effects on reproductive neuroendocine systems, focusing on effects of exposure during fetal development, a life stage that is particularly susceptible to endocrine disruption. We also provide the example of how the use of a whole genome microarray (Affymetrix rat 2.0) to assay gene expression in the preoptic area (POA; a part of the hypothalamus involved in the control of reproductive physiology and behavior) of female rats fetally exposed to PCBs enabled us to determine whether there was wholesale reprogramming of genes in a manner maintained in adulthood. We also use this method to interrogate the pathways by which PCBs exert these effects, and to ascertain any commonalities of potential dose–response relationships, as endocrine-disrupting chemicals (EDCs) often exert non-traditional, non-monotonic effects on physiological systems. The results show that there are indeed a large number of POA genes and pathways perturbed by prenatal PCB exposure. Although some predicted estrogenic pathways were identified by this method, we also identified other hormonal, neurotransmitter, immune, blood, transcriptional, and intracellular signaling pathways that were affected by the prenatal PCBs. Furthermore, this analysis enabled us to show that the dose–response relationships between exposure and gene expression were almost

A.C. Gore (✉)
The Institute for Neuroscience, The University of Texas at Austin, Austin, TX, USA

Division of Pharmacology & Toxicology, College of Pharmacy, The University of Texas at Austin, Austin, TX, USA

Institute for Cellular and Molecular Biology, The University of Texas at Austin, Austin, TX, USA
e-mail: andrea.gore@mail.utexas.edu

J.-P. Bourguignon et al. (eds.), *Multi-System Endocrine Disruption*,
Research and Perspectives in Endocrine Interactions,
DOI 10.1007/978-3-642-22775-2_1, © Springer-Verlag Berlin Heidelberg 2011

entirely non-monotonic, with U-shaped and inverted U-shaped the most common. The example of PCBs as neuroendocrine disruptors may predict the effects of many other compounds, including anthropogenic chemicals used today such as bisphenol A and phthalates.

Introduction to PCBs and their Endocrine-Disrupting Effects

Polychlorinated biphenyls (PCBs) were in common use worldwide in the mid-twentieth century in industrial manufacturing processes. Their production was halted in the United States in 1978 in response to growing public recognition of their toxic effects on the health of humans and wildlife. However, PCBs are still detected worldwide in soil, air, water and the biomass, due to their dispersal, persistence, and the lack of biological mechanisms to degrade, detoxify or eliminate PCBs from the body burden. Since the time of the initial discoveries of PCBs' overt toxicity, it has become clear that lower-dose, sub-toxic exposures to PCBs can cause endocrine-disrupting effects on reproductive, thyroid, and other endocrine systems (Dickerson and Gore 2007).

Developmental exposures to PCBs, particularly to the fetus or infant, are particularly detrimental due to the vulnerability of rapidly growing and dividing cells. This concept of sensitive or critical developmental periods (Barker et al. 2002; Barker 2003) takes into consideration that the fetus/infant is exposed to the external environment via placental or lactational transfer of natural and man-made compounds from the mother's body. Although potentially subtoxic, the cumulative consequence of these exposures on the fetus, together with the exposed individual's own genetic traits, may predispose that individual to disease later in life. The best example of the "fetal basis of adult disease" in humans come from the study of diethylstilbestrol (DES), a pharmaceutical estrogen given to pregnant women to reduce the risk of miscarriage (Herbst et al. 1971). It was noted that there was increased incidence of adenocarcinoma of the vagina in young women, a population that does not normally have this type of cancer, and it was determined retrospectively that these women had been fetally exposed to DES taken by their pregnant mothers. Thus, a direct link between fetal exposure and adult disease was drawn for the first time in humans. Furthermore, a similar phenotype could be produced in animal models of perinatal DES exposure (McLachlan et al. 1982), a finding that was important both because it enabled mechanistic studies of how fetal DES exposure caused this latent disease and because it showed the conservation of endocrine disrupted traits between animal models and humans.

A growing body of literature now shows that low-dose developmental EDC exposures, including PCBs, may permanently reprogram endocrine and reproductive dysfunctions in adulthood (Dumesic et al. 2007; Steinberg et al. 2007, 2008; Diamanti-Kandarakis et al. 2009). As used here, "reprogramming" refers to how environmental changes early in life may change the capacity of a gene or protein to be expressed later in life, even in the absence of further exposure to the causal environmental agent. In the case of PCBs, transient fetal or early postnatal exposures in animal models have been linked to reproductive and thyroid disorders in adulthood (reviewed in Dickerson and Gore 2007). In humans, the cause-and-effect relationship between PCBs and disease is less clear, except for cases of toxic poisoning, such as contaminated cooking oil in Japan and Taiwan (Seegal 1996). Epidemiological studies on humans show correlations, albeit not necessarily direct, between PCB body burden and lower IQ, cognitive dysfunctions, other neurobehavioral abnormalities and, more recently, cardiovascular disease (Winneke et al. 1998; Lang et al. 2008). Although the links between developmental exposure and later life dysfunctions are more difficult to prove, there is no doubt that PCBs continue to serve as a valuable model for understanding the effects of both historical (DES) and modern putative EDCs, such as bisphenol A, phthalates, vinclozolin, and many others, due to the similarity of the mechanisms by which EDCs act.

PCB Actions on Neuroendocrine Systems

Much of the published literature on EDCs has focused on their detrimental effects on reproductive systems. It is important to point out that the control of reproduction goes far beyond the gonads and the reproductive tract, the target of most studies in the past. As is the case for several endocrine systems, reproductive function is much more complex: it involves a series of organs and hormones including the brain, and more specifically the hypothalamus at its base, which produces the neuropeptide gonadotropin-releasing hormone (GnRH); the anterior pituitary, and those cells (gonadotropes) that synthesize and release the gonadotropins, luteinizing hormone (LH) and follicle-stimulating hormone (FSH); and the gonads (testis/ovary), which not only produce germ cells (sperm/ova) but also steroid hormones, such as estrogens, androgens, and progestins, and protein hormones, such as inhibin (reviewed in Gore and Crews 2009; Gore 2010). Of paramount importance is that these hormones regulate one another by feed-forward or feedback mechanisms. GnRH release from the hypothalamus stimulates pituitary LH and FSH release, which in turn stimulate gonadal gametogenesis and steroidogenesis. Sex steroid hormones in the circulation feed back to the hypothalamus and pituitary to regulate GnRH and gonadotropin levels in response to the current need of the organism. This concept is crucial to our understanding of EDC effects, because EDCs can act upon the same hormone receptors, enzyme systems, and other targets in the hypothalamus-pituitary that our endogenous hormones utilize. The remainder of this article will discuss evidence for neuroendocrine disruption by PCBs and will focus on the hypothalamic genes that may be reprogrammed by fetal PCB exposures.

Mechanisms and Consequences of Developmental PCB Action in the Hypothalamus

PCBs are grouped into three main categories: coplanar, dioxin-like coplanar, and non-coplanar, with differential actions on various receptor systems (Dickerson et al. 2009). Most research on endocrine disruption has focused on the estrogenic properties of PCBs and their subsequent actions on estrogen receptors in peripheral reproductive tissues (reviewed in Dickerson and Gore 2007). Furthermore, PCBs can act directly upon neuroendocrine tissues, including hypothalamic neurons and cell lines, in which PCBs alter gene and protein expression, including estrogen receptors and the GnRH peptide (Gore et al. 2002; Salama et al. 2003; Dickerson et al. 2009). The cellular/molecular actions of developmental PCBs in vivo are reflected as differences in reproductive physiology and behavior in adulthood (Chung and Clemens 1999; Chung et al. 2001; Steinberg et al. 2007, 2008), showing that there is reprogramming of genes/proteins that underlie these biological processes.

Research in our laboratory has primarily focused on estrogenic PCBs (Petit et al. 1997; Shekhar et al. 1997; Layton et al. 2002; Ptak et al. 2005) such as Aroclor 1221 (A1221), a lightly-chlorinated commercial PCB mixture composed mostly of mono-ortho-substituted and co-planar congeners (Frame 1997). A1221 binds to estrogen receptor alpha and beta, albeit at relatively low affinity compared to endogenous estrogens (Shekhar et al. 1997; Layton et al. 2002). In addition, A1221 or its constituent congeners can also suppress the ability of the P450 aromatase enzyme to convert testosterone to estradiol (Woodhouse and Cooke 2004; Ptak et al. 2006), thereby affecting local estrogen levels in a negative manner. However, A1221 is not purely estrogenic: exposure to A1221 can mimic thyrotoxicosis and alter circulating thyroid hormone levels (Kilic et al. 2005). One report showed that A1221 could inhibit effects of dihydrotestosterone (an androgen) in a prostate cell line (Schrader and Cooke 2003). Clearly more information is needed, as A1221 cannot be easily classified as an endocrine disruptor of a single class of hormone receptors.

Here, we will summarize the literature on the effects of developmental A1221 exposures on adult neuroendocrine function, as well as studies in the hypothalamic GnRH GT1-7 cell line. Research in this arena began with the reports from Clemens' group that perinatal exposure to PCBs caused changes in feminine sexual behavior in rats (Chung and Clemens 1999; Chung et al. 2001; Wang et al. 2002). Based on this research, we began studying the effects of low dose exposure to A1221 given to pregnant rats during late gestation, which is part of the critical period of brain sexual differentiation in the developing fetus. This work has been published and reviewed extensively (Gore 2008, 2010), so we will provide a relatively short discussion, to be followed by a more in-depth presentation of new data on the effects of developmental exposure to A1221 on gene expression in the preoptic area of the hypothalamus.

In vitro studies were performed to examine the effects of A1221 on GnRH gene expression and cell death in immortalized GnRH GT1-7 cell cultures (Gore et al. 2002). When A1221 was given in dose–response experiments, GnRH mRNA levels and GnRH peptide levels were elevated, particularly at intermediate dosages (Gore et al. 2002). Some (but not all) of these effects were blocked by the estrogen receptor antagonist ICI 182,780, suggesting that there were estrogen receptor-mediated properties of A1221 on GnRH function but that other non-estrogenic mechanisms are probably also involved. We followed up on this work by interrogating the mechanism by which PCBs might cause toxicity in this neuroendocrine cell line. Although in our second study we did not specifically test A1221, we found that individual PCB congeners stimulated/inhibited GnRH gene expression depending upon dose and duration, with stimulatory effects caused at lower doses/shorter time points and inhibitory effects at the higher doses/longer time points (Dickerson et al. 2009). We also observed some reversibility of these effects by the estrogen receptor antagonist, supporting our work on A1221 described above. Finally, we found that GT1-7 cell death caused by the PCBs comprised a mix of apoptotic and necrotic mechanisms, again dependent upon dose/duration of the PCB treatment (Dickerson et al. 2009). These studies as a whole support the potential direct effects of PCBs on GnRH neurons, at least in vitro.

In vivo studies have focused on the effects of developmental A1221 on adult reproductive physiology, behavior, and protein expression. In a series of studies, we showed a decrease in the number of cells expressing estrogen receptor beta (ERβ) protein in a sub-region of the hypothalamus, the anteroventral periventricular nucleus (AVPV), of the rats treated perinatally with A1221 (Salama et al. 2003). We went on to demonstrate that developmental A1221 exposure impaired paced mating behaviors in the female offspring when tested in early adulthood (Steinberg et al. 2007). In that study, female rats were exposed prenatally to vehicle (DMSO) or one of three doses (0.1, 1, 10 mg/kg) of A1221 on embryonic days 16 and 18, part of the "critical period" of brain sexual differentiation. The paced mating paradigm was used because it enables a focus on female-typical behaviors, as the female rat (not the male) controls the pace of mating, and the female can choose whether (or not) to mate and can also control the timing of her contact with the male. Although not all aspects of mating behaviors were affected, we found the following significant, dose-dependent changes in rats exposed prenatal A1221: the amount of time a female rat took to return to a male rat after he had ejaculated was increased (at the 1 mg/kg dose); an increased number of mating trials were required for the female to mate (at the 1 and 10 mg/kg doses); and a decrease in audible vocalizations (1 mg/kg) were detected (Steinberg et al. 2007). We found it particularly interesting that the 1 mg/kg (intermediate) dosage had the greatest effect upon the behavioral phenotype, whereas the low (0.1 mg/kg) and the high (10 mg/kg) dosages were less consistently effective. This finding will be brought to bear in our gene expression studies discussed later in this article and is relevant to the concept that both endogenous hormones and xenobiotics can exert non-monotonic dose–response effects.

The female siblings of the rats used in the behavioral tests were subjected to other analyses of reproductive physiology, both in the fetally exposed (F1 generation) females and also in their female offspring (F2 generation; Steinberg et al. 2008). In the F1 females studied 1 day following a mating trial, serum LH was elevated in the 1 mg/kg group. A more pronounced phenotype was found in the F2 generation, which differed from the F1s in that the F2s were virgin rats and were studied across the estrous cycle. For the F2 generation, serum progesterone and LH and uterine and ovarian weights exhibited predicatable fluctuations across the estrous cycle in the vehicle (control)-descendant rats. By contrast, the A1221-descendant F2 females had abnormal cycles of these serum hormones and reproductive tissue weights, in a manner suggestive of a disruption of the circadian rhythms of these processes (Steinberg et al. 2008).

Current research in the laboratory is focused on the effects of A1221 on *three* generations of male and female offspring. We are also evaluating the molecular consequences of the prenatal exposure, beginning with the F1 males and females. Most recently, we have observed that the sexual differentiation of the estrogen receptor alpha (ERα) protein in the preoptic area (POA) is perturbed, an effect seen as early as the day after birth (postnatal day 1; Dickerson et al. 2011b), and is maintained into adulthood (Dickerson et al. 2011). We have also seen perturbations of GnRH neural activation, as indicated by co-expression of the immediate early gene product Fos, and disruptions of expression of the neuropeptide kisspeptin (Dickerson et al. 2011), an important regulator of GnRH function (Smith et al. 2005).

Hypothalamic Gene Expression Reprogramming by PCBs

In the remainder of this article, we will present and discuss data from analyses of whole genome effects of prenatal A1221 exposure on gene expression in the POA, a neuroendocrine control center that forms a functional unit with the hypothalamus proper. We had three overarching hypotheses. First, we wished to determine whether there was wholesale reprogramming of gene expression caused by fetal PCB exposure in a manner that was maintained in adulthood. Although the published literature on the “fetal basis of adult disease” shows such long-term latent effects, we felt that it was important to confirm this finding on a broad scale, and a whole genome microarray analysis (Affymetrix) would reveal these effects. Second, we wanted to confirm our behavioral and hormonal observations that intermediate dosages of PCBs were most effective in causing an endocrine disruption. This hypothesis is based on a broader literature showing non-linear dose–response curves, such as inverted-U or U-shaped curves, for effects of EDCs or endogenous hormones (Calabrese and Baldwin 2001; Sheehan 2006). By examining gene expression in our animals from the dose–response study, we can infer whether A1221 is affecting transcription via a non-monotonic steroid hormone-like pathway, or via another mechanism. Third, we could use the microarrays to challenge the dogma that PCBs such as A1221 are primarily estrogenic in their actions. As discussed earlier, A1221 can bind to estrogen receptor α (ERα) and ERβ

(Shekhar et al. 1997; Layton et al. 2002), and its effects can be blocked by ER antagonists, such as ICI 182,780 (Gore et al. 2002; Dickerson et al. 2009). Nevertheless, A1221 also interferes with the ability of the aromatase to convert testosterone to estradiol (Woodhouse and Cooke 2004; Ptak et al. 2006), and it has anti-androgenic (Schrader and Cooke 2003) and thyroid (Kilic et al. 2005) properties. The field of endocrine disruption has been overly simplistic in trying to attribute effects of a single endocrine disruptor (or a mix of PCBs such as A1221) to a single receptor-mediated mechanism. Therefore, the next section will provide evidence to address these three hypotheses.

Microarray Study on Effects of PCBs on Gene Expression in the POA

The POA is a key sexually dimorphic brain region that is integrally involved in the control of reproductive physiology and behavior (Raisman and Field 1971; Tobet and Hanna 1997; Scott et al. 2000). In this experiment, we utilized the POAs from F1 females used in our published studies (Steinberg et al. 2007, 2008), euthanized the day after a paced mating experiment. The POA was dissected ($n = 5, 4, 5, 4$ females in the control, 0.1, 1, and 10 mg/kg treatment groups, respectively) and RNA was extracted following published methods (Walker et al. 2009). RNA was amplified by Ambion, Inc. (Austin, TX), using the MessageAmp II Biotin kit (Ambion #1791), and amplification products were assessed using the Agilent Bioanalyzer 2100. Hybridization of samples to the rat 230 2.0 BioChip and image capture were carried out according to Affymetrix protocols, and the original Affymetrix CDF file was used for annotation and processing of microarray expression values. The Affymetrix Statistical Algorithm software GCOSv1.3 was used to quantify image signal results. Because samples were run in two separate batches, Batch Model Statistics were used to account for batch effects. We used the R software environment and a mixed linear model ANOVA with the Maanova package for analyzing the expression data (Kerr et al. 2000). Raw data were compiled as expression values using the RMA (Robust Multichip Average) expression measure for Affymetrix GeneChip® arrays (Bolstad et al. 2003).

Gene Expression Transcript Changes Caused by Prenatal PCB Treatment

ANOVA results revealed 1,512 transcripts with a nominal $p < 0.05$ and 405 transcripts having a nominal $p < 0.01$. Due to the large number of tests and potential corrections reducing statistical power, data shown here are raw (uncorrected) p-values. The number of genes from this list were grouped according to chromosomal location and normalized to the overall number of genes on the microarray belonging to the same chromosome. No bias was observed in the chromosomal location of affected genes (3.5–6.7% of genes on each chromosome were represented). The 215 genes with proposed functions are listed in Table 1, grouped categorically by p-value. We point out that follow-up analysis by quantitative PCR or other methods is necessary to confirm the microarray results and that the results would be strengthened by increasing the statistical power.

Predicting that the POA genes most strongly affected by PCB exposure would be hormone-related or -regulated (e.g., estrogen, thyroid, retinoid, and stress) and neurotransmitter-related genes, and together with guidance from the literature on in vivo and in vitro effects of PCBs, we selected 26 genes for further *post-hoc* analysis (Table 2). For example, both ERs (ERα and ERβ) were identified, as were a number of genes involved in estrogen signaling in breast cancer, such as MapK15/ERK8, “similar to breast cancer membrane protein 101,” “similar to breast cancer associated oncogene”/BCA3, “secretoglobin family 3A member 1”/HIN1, OKL38 and mac25. Other nuclear receptor family members and enzymes involved in thyroid hormone and metabolic function were identified, including the thyroid hormone receptor associated protein 1, “stimulated by retinoic acid gene 6 homolog” and retinol dehydrogenase 2.

Stress pathways have not been widely studied in endocrine disruption, but we found several potential candidates from our microarrays. The expression of corticotropin releasing hormone binding protein (CRHBP), corticotropin releasing hormone receptor 1 (CRHR1), CRHR2 and nuclear receptor coactivator 6 (involved in stress as well as other endocrine pathways; Tetel 2009) differed among the treatment groups. Previously, methylsulfonyl PCB metabolites had been reported to alter corticosterone synthesis (Johansson et al. 1998a) and bind to the glucocorticoid receptor (Johansson et al. 1998b), and PCB exposure is linked with elevated basal corticosterone levels (Miller et al. 1993). Our study presents the first evidence, to our knowledge, of a PCB effect on CRHBP expression, along with “14-3-3 protein”/“tyrosine 3-monooxygenase/tryptophan 5-monooxygenase activation protein” (a glucocorticoid receptor signal transduction regulator), and CRHR1 and CRHR2.

Neural and glial markers, growth factors, and neurotransmitters were also identified (Tables 1 and 2). Both neuregulin 1 and neuregulin 2, cytokines for which mutation or altered expression is related to neuropathologies, including Alzheimer’s and schizophrenia susceptibility (Esper et al. 2006; Meeks et al. 2006), were included amongst the most significantly affected genes. In addition, spondin 1 is known to interact with Alzheimer’s-related protein amyloid beta precursor protein (APP) to inhibit its cleavage (Ho and Sudhof 2004). Environmental toxicant exposure may be linked to neurodegenerative diseases and dementia (Troster et al. 1991; Corrigan et al. 2000; Landrigan et al. 2005), and the gene expression results support several candidates for future analysis.

As a whole, the gene expression results revealed a number of POA genes affected by prenatal PCB exposure, several of which are involved in neural and endocrine functions. It should be noted that most genes, however, were in other

Table 1 Gene transcripts identified by microarray analysis in the adult POA of rats treated prenatally with PCBs or vehicle

Nominal $p < 0.0005$	
Hemoglobin beta chain complex (appears twice)	Lumican
Similar to heterogeneous nuclear ribonucleoprotein G – human	T-box 2 (predicted)
Hemoglobin alpha, adult chain 1 or alpha 2 chain (appears twice)	Cathepsin L
Glucocorticoid modulatory element binding protein 2	Opioid binding protein/cell adhesion molecule-like
C1q and tumor necrosis factor related protein 4 (predicted)	Similar to inositol 1,3,4-triphosphate 5/6 kinase
Similar to NADH dehydrogenase (ubiquinone) 1, subcomplex unknown, 1 (predicted)	F-box only protein 32
MAM domain containing glycosylphosphatidylinositol anchor 1 (predicted)	Churchill domain containing 1 (predicted)
Cytochrome P450, family 7, subfamily a, polypeptide 1	Gap junction membrane channel protein alpha 5
Similar to unconventional myosin-9b	
Similar to spermatogenesis associated glutamate (E)-rich protein 4d	
(Nominal p-value: $0.0005 < p < 0.005$)	
Kruppel-like factor 12 (predicted)	Similar to C11orf17 protein (predicted)
Similar to breast cancer membrane protein 101 (predicted)	CD5 antigen-like
Phospholipase A2, group IIA (platelets, synovial fluid)	Insulin receptor-related receptor
Talin 1	Upstream transcription factor 2
Mitogen-activated protein kinase 15	Prohibitin 2
Crystallin, gamma S	XK-related protein 5
Similar to CREBBP/EP300 inhibitory protein 1 (predicted)	Ubiquitin D
Intraflagellar transport 74 homolog (Chlamydomonas)	Aminolevulinic acid synthase 2
Extracellular link domain-containing 1 (predicted)	RT1 class Ib, locus S3
Solute carrier family 22 (organic cation transporter), member 2	ADP-ribosyltransferase 5
Stimulated by retinoic acid gene 6 homolog (mouse)	Cut-like 2 (Drosophila) (predicted)
ELAV (embryonic lethal, abnormal vision, Drosophila)-like 1 (Hu antigen R) (predicted) (appears twice)	Selenium binding protein 2
Enthoprotin	Neuregulin 1
Similar to [Ascaris lumbricoides mRNA sequence], gene product (predicted)	Osteoglycin (predicted)
FK506 binding protein 9	Syntaxin binding protein 1
Reproductive homeobox on X chromosome, 9	Ankyrin 3, epithelial
Integrin beta 3 binding protein (beta 3-endonexin)	Triadin
Reproductive homeobox on X chromosome, 9	Splicing factor 3a, subunit 1 (predicted)
Nicotinamide N-methyltransferase (predicted)	Similar to C14orf25 protein (predicted)
Gap junction membrane channel protein beta 3	Syntaxin 8

(continued)

Table 1 (continued)

Cyclin M1 (predicted)	2,3-bisphosphoglycerate mutase
Nudix (nucleotide diphosphate linked moiety X)-type motif 3	Casein kinase II, alpha 1 polypeptide
SH2 domain binding protein 1 (tetratricopeptide repeat containing)	Chloride channel K1
Mitogen-activated protein kinase 8 interacting protein 3	Guanine nucleotide binding protein, alpha 14
G protein-coupled receptor 39	ADP-ribosyltransferase 1 (predicted)
Similar to Zinc finger protein 184 (predicted)	Notch gene homolog 2 (Drosophila)
Low density lipoprotein receptor-related protein 1	Vimentin
Retinoic acid-induced 1 (predicted)	Similar to C11orf17 protein (predicted)
Insulin-like growth factor binding protein 7	C-reactive protein, pentraxin-related
V-ets erythroblastosis virus E26 oncogene like (avian)	Adenylate cyclase 6
Cytochrome P450, subfamily 2G, polypeptide 1	Insulin-related protein 2 (islet 2)
Similar to neurobeachin (predicted)	CD24 antigen
Similar to WAC (predicted)	DEAD (Asp-Glu-Ala-Asp) box polypeptide 19
Melanocortin 5 receptor	Corticotropin releasing hormone receptor 2
Kinesin light chain 3	Similar to Ras-related protein Rab-1B
Laminin, alpha 2 (predicted)	Upstream binding protein 1 (predicted)
Complement component 1, q subcomponent, gamma polypeptide	Similar to cDNA sequence BC020002 (predicted)
Glutamyl aminopeptidase	Homer homolog 1 (Drosophila)
Prolylcarboxypeptidase (angiotensinase C) (predicted)	Unknown (protein for MGC:72614)
Smg-7 homolog, nonsense mediated mRNA decay factor (*C. elegans*) (predicted)	Noggin
Natriuretic peptide precursor type C	Similar to nemo-like kinase (predicted)
Hairy/enhancer-of-split related with YRPW motif-like (predicted)	Grainyhead-like 1 (Drosophila) (predicted)
AHNAK nucleoprotein (desmoyokin)	Zinc finger protein 503 (predicted)
Chondroitin sulfate proteoglycan 4	Tocopherol (alpha) transfer protein
Thyroid hormone receptor associated protein 1 (predicted)	Nuclear receptor coactivator 6
CCAAT/enhancer binding protein , epsilon	Glutathione S-transferase A3
Chemokine (C-C motif) ligand 22	Osteoglycin (predicted)
Interferon-induced transmembrane protein 1 (predicted)	Similar to AP2-associated kinase 1 (predicted)
SAM domain, SH3 domain and nuclear localization signals, 1	Similar to ovostatin-2 (predicted)
Interleukin 2 receptor, gamma (severe combined immunodeficiency)	Protein kinase C, alpha

(continued)

Table 1 (continued)

PRKC, apoptosis, WT1, regulator	Gap junction membrane channel protein beta 2
Cyclin-dependent kinase (CDC2-like) 10	
Nominal p-value: 0.005 < p < 0.01	
Potassium inwardly-rectifying channel, subfamily J, member 5	Solute carrier family 9, member 1
Adrenergic receptor, alpha 2b	Neurobeachin-like 2 (predicted)
Paired-like homeodomain transcription factor 1	Interleukin 24
Predicted: similar to lipocalin-interacting membrane receptor	Ciliary neurotrophic factor
Gastric inhibitory polypeptide	Neuregulin 2
X transporter protein 3	Annexin A1
Hydroxysteroid (17-beta) dehydrogenase 9	Ribosomal protein L17
Similar to brain carcinoembryonic antigen	Pregnancy-induced growth inhibitor
Anterior pharynx defective 1b homolog (*C. elegans*)	Ribonucleotide reductase M1 (mapped)
Predicted: similar to PHD finger protein 20-like 1 isoform1	Fibrinogen, alpha polypeptide
Coatomer protein complex, subunit beta 2 (beta prime)	Melanoma cell adhesion molecule
Proteosome (prosome, macropain) subunit, beta type 9	Allograft inflammatory factor 1
Zinc finger protein 213 (predicted)	Cyclin G2 (predicted)
Beta heavy chain of outer-arm axonemal dynein ATPase	Defensin beta 1
Similar to replication protein-binding trans-activator RBT1	Chemokine (C-C motif) ligand 4
G protein-coupled receptor 23 (predicted)	Fibroblast growth factor 21
Mitochondrial tumor suppressor 1	Interferon beta 1, fibroblast
Centaurin, beta 5 (predicted)	Troponin I type 3 (cardiac)
Toll interacting protein (predicted)	Reticulocalbin 1 (predicted)
Spondin 1	Zinc finger, FYVE domain containing 9 (predicted)
Citrate lyase beta like	Complement component 3
Deiodinase, iodothyronine, type II	Crystallin, beta B3
Similar to mKIAA0159 protein (predicted)	Zinc finger protein 96
Similar to RNA (guanine-9-) methyltransferase domain containing 2	Valyl-tRNA synthetase 2
Moloney leukemia virus 10-like 1 (predicted)	Secretin
Potassium inwardly-rectifying channel, subfamily J, member 3	Paraoxonase 1
Nudix (nucleoside diphosphate linked moiety X)-type motif 6	Cathepsin W
Procollagen, type VI, alpha 3 (predicted)	Ras homolog gene family, member V
Calcium channel, voltage-dependent, L type, alpha 1 C subunit	Transgelin 2
Similar to microtubule associated serine/threonine kinase 2 (predicted)	RT1 class I, CE16
Polymerase (RNA) II (DNA directed) polypeptide H (predicted)	Pbx/knotted 1 homeobox
Similar to RAP2A, member of RAS oncogene family (predicted)	Similar to Ab2-095
Ring finger protein 190	Roundabout homolog 2 (Drosophila)
Myeloid/lymphoid or mixed-lineage leukemia (trithorax homolog, Drosophila); translocated to, 11	Similar to Protein C20orf129 homolog (predicted)

(continued)

Table 1 (continued)

Insulin-responsive sequence DNA binding protein-1	Similar to IQ motif and WD repeats 1 (predicted)
Regulator of nonsense transcripts 1 (predicted)	Myosin IG
DNA cross-link repair 1B, PSO2 homolog (*S. cerevisiae*)	Similar to KIAA0339 protein
Phospholipase A2, group IVA (cytosolic, calcium-dependent)	Phospholipase C, beta 4
Tyrosine 3-monooxygenase/tryptophan 5-monooxygenase activation protein, eta polypeptide	Vacuolar protein sorting 37 C (yeast) (predicted)
Solute carrier family 39 (iron-regulated transporter), member 1	Interleukin 17B
Purinergic receptor P2X, ligand-gated ion channel, 5	Regulator of G-protein signaling 6
RAB5B, member RAS oncogene family (predicted)	Nephroblastoma overexpressed gene
McKusick-Kaufman syndrome protein	Secretoglobin, family 3A, member 1

Transcripts with no assigned name (ESTs or unknown genes) and multiple transcripts from the same gene are not included in this list, resulting in 215 genes. These are grouped by nominal p-values: $p < 0.0005$, $0.0005 < p < 0.005$, and $0.005 < p < 0.01$. Data are shown in 2 columns.

functional groups, as discussed below. The results highlight that a variety of physiological pathways – not just the traditional estrogenic pathways normally postulated to underlie the effects of lightly chlorinated PCBs such as A1221 – are affected in adulthood following prenatal PCB exposure.

Functional Analyses of Identified Genes

Gene Ontology (GO) functional grouping analysis was used to group the identified genes by their biological roles in the nervous system (Table 3). In total, 48% of transcripts significantly affected by treatment had no known identity or function. Genes from this list were each researched using published studies and databases and were categorized according to protein function in a custom functional analysis (Fig. 1). Results by the GO method and our custom pathway analyses were in agreement. Interestingly, the top five categories were blood/immune, transcription factors, cytokines and neurotransmission, intracellular signaling, and hormone signaling. Functional group showed that fewer than 10% of the identified genes were involved in hormone signaling.

We were surprised by the results of the analysis showing that the primary categories of significantly affected genes were blood and immune-related. PCB exposure can decrease erythrocyte count and platelet volume, as well as lower immune response, and alter ratios of T helper and suppressor cells (Arnold et al. 1993). In addition, PCBs can bind to hemoglobin directly (Rehulka and Minarik 2004), which may lead to premature degradation. Thus, links between early life PCB exposure and neuroimmunology merit further study.

Table 2 Candidate gene list

Affymetrix ID	Nominal p-value	Gene ID	Significant post-hoc expression differences with control ($p < 0.05$)
1381606_at	0.019	Stimulated by retinoic acid gene 6 homolog	
1371142_at	0.003	Cytochrome P450, subfamily 2 G, polypeptide 1	
1389931_at	0.004	Corticotropin releasing hormone receptor 2	1 mg/kg < Control
1372176_at	0.004	Protein kinase C, alpha	
1371072_at	0.005	Nuclear receptor coactivator 6	1 mg/kg < Control
1386156_at	0.005	Thyroid hormone receptor associated protein 1	1 mg/kg > Control
1370567_at	0.005	Adrenergic receptor, alpha 2b	
1369898_a_at	0.005	Gastric inhibitory polypeptide	
1387994_at	0.005	Hydroxysteroid (17-beta) dehydrogenase 9	1 and 10 mg/kg < Control
1391033_s_at	0.007	Neuregulin 2	
1369908_at	0.011	Corticotropin releasing hormone binding protein	1 and 10 mg/kg < Control
1381449_s_at	0.011	Transforming growth factor alpha	
1370566_at	0.015	Retinol dehydrogenase 2	
1369533_a_at	0.016	5-hydroxytryptamine (serotonin) receptor 4	0.1 mg/kg < Control
1384868_at	0.024	Estrogen receptor alpha	10 mg/kg > Control
1388260_a_at	0.024	Leptin receptor	
1398289_a_at	0.025	Corticotropin releasing hormone receptor 1	0.1 mg/kg < Control
1388009_at	0.030	Thyrotropin releasing hormone receptor 2	10 mg/kg > Control
1367851_at	0.030	Prostaglandin D2 synthase	
1371762_at	0.036	Retinol binding protein 4, plasma	
1368285_at	0.037	Sex hormone binding globulin	
1369760_a_at	0.038	Estrogen receptor beta	10 mg/kg > Control
1388239_at	0.038	Period homolog 3	
1382916_at	0.038	Thyroid hormone receptor beta	
1371120_s_at	0.041	Bradykinin receptor, beta 2	
1375720_at	0.043	Gamma-aminobutyric acid (GABA) B receptor 1	

Tukey-Kramer post-hoc analysis was performed on 26 selected genes, with directionality of the difference between PCB- and DMSO-treated rats at a significance level of $p < 0.05$

Table 3 Gene Ontology (GO) results for biological processes

(1) Regulation of biological process (15)
- (A) Regulation of enzyme activity (1)
  - (i) Positive regulation of enzyme activity
- (B) Regulation of physiological process (8)
  - (i) Regulation of secretion (1)
  - (ii) Regulation of vascular permeability (1)
  - (iii) Regulation of metabolism (6)
- (C) Regulation of development (5)
  - (i) Regulation of growth (3)
  - (ii) Regulation of cell differentiation (2)
- (D) Regulation of cellular process (7)
  - (i) Regulation of cell proliferation (4)
  - (ii) Regulation of cell size (4)
  - (iii) Regulation of cell volume (1)
  - (iv) Regulation of membrane potential (1)
  - (v) Regulation of programmed cell death (2)
  - (vi) Regulation of growth (3)
  - (vii) Regulation of cell differentiation (2)
  - (viii) Regulation of signal transduction (2)

(2) Physiological process (54)
- (A) Secretion (3)
  - (i) Regulation of secretion (1)
  - (ii) Fluid secretion (1)
  - (iii) Hormone secretion (2)
- (B) Regulation of physiological process (8)
  - (i) Regulation of secretion (1)
  - (ii) Regulation of vascular permeability (1)
  - (iii) Regulation of metabolism (6)
- (C) Response to stimulus (13)
  - (i) Response to stress (7)
  - (ii) Response to external stimulus (12)
  - (iii) Response to endogenous stimulus (1)
- (D) Organismal physiological process (7)
  - (i) Regulation of body fluids (2)
  - (ii) Circulation (1)
  - (iii) Excretion (1)
  - (iv) Immune response (5)
- (E) Coagulation (1)
  - (i) Blood coagulation (1)
- (F) Death (2)
  - (i) Cell death (2)
- (G) Homeostasis (2)
  - (i) Ion homeostasis (1)
  - (ii) Cell homeostasis (2)
- (H) Metabolism (28)
  - (i) Regulation of metabolism (6)
  - (ii) Protein metabolism (12)
  - (iii) Organic acid metabolism (2)

(continued)

Table 3 (continued)

(iv) Oxygen and reactive oxygen species metabolism (1)
(v) Phosphorus metabolism (4)
(vi) Nucleobase, nucleoside, nucleotide and nucleic acid metabolism (8)
(vii) Energy pathways (1)
(viii) Hormone metabolism (2)
(ix) Lipid metabolism (5)
(x) Catabolism (6)
(xi) Carbohydrate metabolism (1)
(xii) Biosynthesis (3)
(xiii) Aromatic compound metabolism (1)
(xiv) Amino acid derivative metabolism (1)
(xv) Amine metabolism (1)
(xvi) Alcohol metabolism (1)
(I) Cellular physiological process (17)
(i) Cell motility (2)
(ii) Cell growth and/or maintenance (15)
(iii) Cell death (2)
(3) Development (13)
(A) Regulation of development (5)
(i) Regulation of growth (3)
(ii) Regulation of cell differentiation (2)
(B) Morphogenesis (11)
(i) Organogenesis (7)
(ii) Appendage morphogenesis (1)
(iii) Embryonic morphogenesis (1)
(iv) Cellular morphogenesis (4)
(C) Growth (6)
(i) Regulation of growth (3)
(ii) Tissue regeneration (2)
(iii) Cell growth (4)
(D) Embryonic development (1)
(i) Embryonic morphogenesis (1)
(E) Cell differentiation (4)
(i) Neuron differentiation (2)
(ii) Regulation of cell differentiation (2)
(iii) Myeloid blood cell differentiation (1)
(F) Cell fate commitment (1)
(i) Cell fate determination (1)
(4) Cellular process (31)
(A) Regulation of cellular process (7)
(i) Regulation of cell proliferation (4)
(ii) Regulation of cell size (4)
(iii) Regulation of cell volume (1)
(iv) Regulation of membrane potential (1)
(v) Regulation of programmed cell death (2)
(vi) Regulation of cell growth (3)
(vii) Regulation of cell differentiation (2)
(viii) Regulation of signal transduction (2)

(continued)

Table 3 (continued)

(B) Cellular physiological process (17)
(i) Cell motility (2)
(ii) Cell growth and/or maintenance (15)
(iii) Cell death (2)
(C) Cell differentiation (4)
(i) Neuron differentiation (2)
(ii) Regulation of cell differentiation (2)
(iii) Myeloid blood cell differentiation (1)
(iv) Cell fate commitment (1)
(D) Cell communication (16)
(i) Signal transduction (11)
(ii) Cell-cell signaling (4)
(iii) Cell adhesion (3)
(5) Behavior (1)
(A) Feeding behavior (1)

Gene Ontology (GO) chart categorizing genes with similar biological processes, with a p-value < 0.01. Results extracted via SwisProt ID of genes, using MPSS software (http://www.scbit.org/mpss/) (Hao et al. 2005). Numbers of genes are shown in parentheses. Additional search engines were used to determine the identity of the proteins: the NetAffx web search engine, BLAST, searches of transcript sequences, and/or RefSeq searches for highly homologous genes in other species (>98.5% homology), and PubMed literature searches were conducted to accurately determine functional roles of each protein product within the brain, and Information Hyperlinked Over Proteins (ihop) and Online Mendelian Inheritance in Man (OMIM) databases were consulted for additional, cross-species information

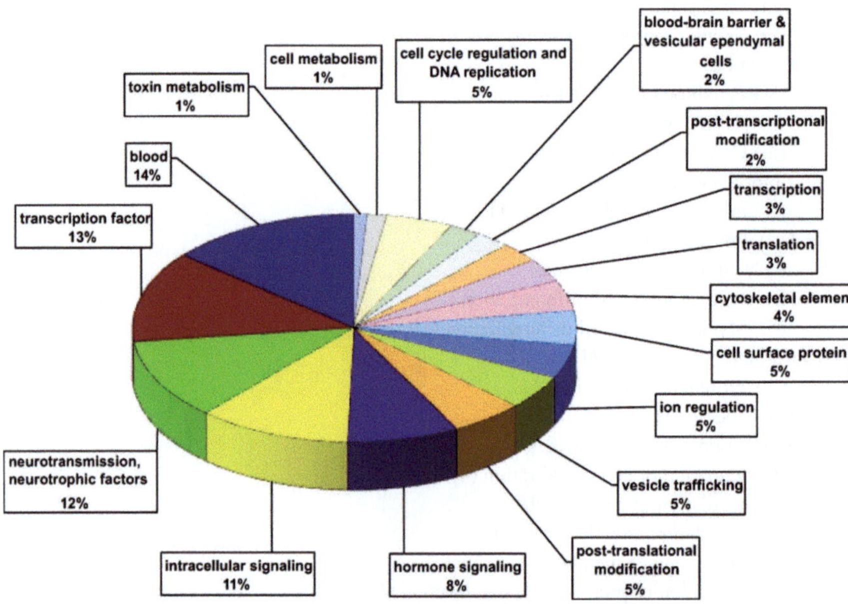

Fig. 1 A custom analysis was performed to group the identified genes categorically into biological functions in the nervous system. The number and percentage of genes in each category are shown

Cluster Analysis of Dose Response Relationships

The field of endocrine disruption has proven contentious in part because dose–response analyses have revealed non-monotonic relationships between exposures and outcome (Diamanti-Kandarakis et al. 2009). Cluster analysis provides a non-subjective approach to grouping data by the shape of the dose–response curves, thus enabling us to assess the relative frequencies of dose–response curves when non-linear relationships are expected, as with U-shaped or inverted U-shaped dose–response curves that often accompany endocrine disrupting chemicals in toxicological research (Weltje et al. 2005).

Of the 405 genes submitted to analysis, 264 were placed within a cluster (~65%). The resulting 11 clusters (Fig. 2) are organized by the shape of the curves. No linear

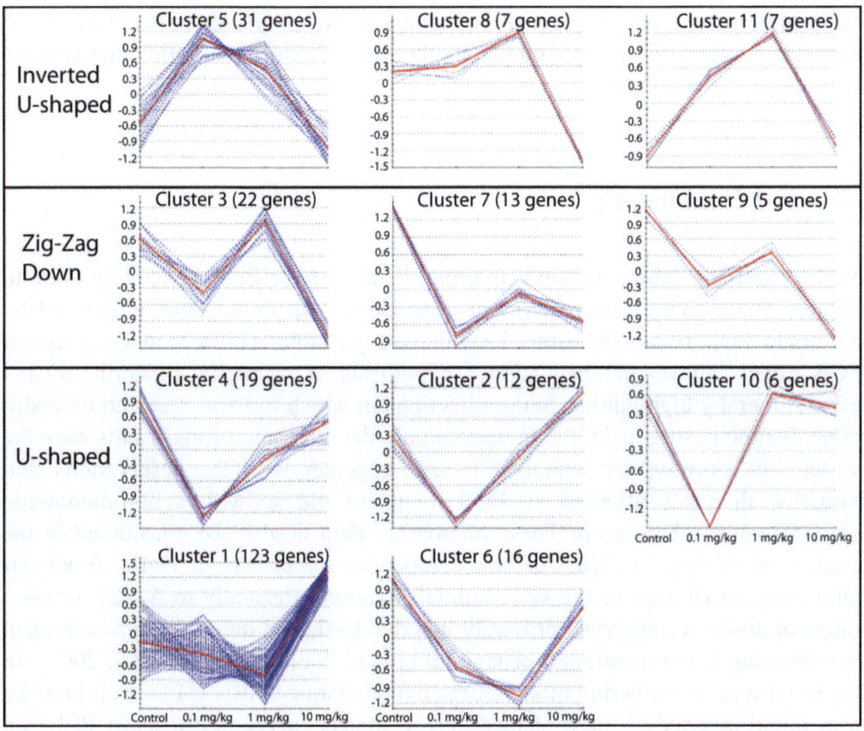

Fig. 2 Cluster analysis results for genes with a p-value <0.01. Data were generated on Adaptive Quality-Based Clustering software (De Smet et al. 2002). For a cluster membership probability of 60% and a minimum of four genes per cluster, 11 graphs were generated, with cluster numbers arbitrarily assigned prior to analysis. These have been grouped according to dose–response curve shape: Inverted U-shaped, Zig-Zag Down, and U-shaped. Data from each transcript are shown as a *blue dashed line*, and averaged curves are given in a *solid red line*. The Y-axis is centered around a normalized mean of all transcript expression values included in the cluster, given as 0. Deviation from this overall mean is indicated by positive or negative deflections from the 0 mark. Dose–response relationships that did not conform to a common cluster are not included in these results

dose–response curves emerged as a significant cluster from the analysis parameters employed in this study. The majority of genes showed U-shaped dose response curves (Fig. 2); 40 genes exhibited a Zig-zag Down effect wherein the 1 mg/kg group differed from other groups; and 45 genes showed an inverted U-shaped dose–response curve (Fig. 2).

U-shaped and inverted U-shaped dose–response curves are often elicited by steroid hormone or xenobiotic actions, including studies on endocrine disruptors in neuroendocrine cells (Weltje et al. 2005). The preponderance of U-shaped dose–response curves suggests that A1221 is acting via a steroid hormone-like mechanism. Possible mechanistic explanations for non-monotonic dose–response curves include actions of endocrine disruptors such as A1221 on steroid hormone receptors, alteration of enzymatic activity related to steroid hormone signaling or metabolism, direct binding to orphan nuclear receptors such as AhR, which may result in steroid hormone-like dose–response relationships, or increased susceptibility of hormone-producing/sensitive regions, particularly such as the POA, which has a leaky blood–brain barrier and may therefore be more susceptible to exposure to circulating xenobiotics.

General Conclusions

We viewed the whole genome microarray as a hypothesis-generating tool to explore broad changes in patterns of gene expression in the POA influenced by environmental toxicant exposure. The microarray enables a broader investigation of PCB effects above and beyond our hypothesis of primarily endocrine-related effects, thereby highlighting future directions in which to focus research on endocrine disruptors such as PCBs. In our opinion, the most interesting results were the variety of pathways by which PCBs act, together with the observations that virtually all genes affected by PCB treatment did so with a non-monotonic dose–response relationship. These microarray data need to be considered in the context of other published work on neuroendocrine effects of PCBs. From our laboratory, in siblings of the same animals exposed prenatally to A1221 across a range of doses, we observed primarily inverted U-shaped dose–response curves in developmental, hormonal, and behavioral effects (Steinberg et al. 2007, 2008). In the in vitro neuroendocrine model of the hypothalamic GnRH GT1-7 cell line, we also found inverted U- or U-shaped dose response curves in effects of PCBs on GnRH release and gene expression (Gore et al. 2002) and on cell death mediated by apoptosis and/or necrosis (Dickerson et al. 2009). The patterns of dose–response curve clusters detected by the current microarray data support these earlier experimental data. Moreover, while it is known that PCBs act by a variety of cellular/molecular mechanisms, too frequently, investigators try to fit these (and other endocrine disruptors) into a simple molecular class, calling them "estrogenic," "anti-androgenic," "thyroid disrupting," etc. The microarray data shown here indicate a range of molecular mechanisms utilized by PCBs even in a relatively specific tissue dissection of the POA. Although we emphasize the preliminary nature of the microarray data that are discussed here for hypothesis-generating purposes, we believe that these data are a valuable resource for future investigation.

Acknowledgments We acknowledge generous support from the PhRMA Foundation (Predoctoral Fellowship to RMS) and the NIH-NIEHS (ES12272, ES07784, ES018139 to ACG).

References

Arnold DL, Bryce F, Karpinski K, Mes J, Fernie S, Tryphonas H, Truelove J, McGuire PF, Burns D, Tanner JR, Stapley R, Zawidzka ZZ, Basford D (1993) Toxicological consequences of Aroclor 1254 ingestion by female rhesus (Macaca mulatta) monkeys. Part 1B. Prebreeding phase: clinical and analytical laboratory findings. Food Chem Toxicol 31:811–824

Barker DJP (2003) The developmental origins of adult disease. Eur J Epidemiol 18:733–736

Barker DJ, Eriksson JG, Forsen T, Osmond C (2002) Fetal origins of adult disease: strength of effects and biological basis. Int J Epidemiol 31:1235–1239

Bolstad BM, Irizarry RA, Astrand M, Speed TP (2003) A comparison of normalization methods for high density oligonucleotide array data based on variance and bias. Bioinformatics 19:185–193

Calabrese EJ, Baldwin LA (2001) Hormesis: a generalizable and unifying hypothesis. Crit Rev Toxicol 31:353–424

Chung Y-W, Clemens LG (1999) Effects of perinatal exposure to polychlorinated biphenyls on development of female sexual behavior. Bull Environ Contam Toxicol 62:664–670

Chung YW, Nunez AA, Clemens LG (2001) Effects of neonatal polychlorinated biphenyl exposure on female sexual behavior. Physiol Behav 74:363–370

Corrigan FM, Wienburg CL, Shore RF, Daniel SE, Mann D (2000) Organochlorine insecticides in substantia nigra in Parkinson's disease. J Toxicol Environ Health A 59:229–234

De Smet F, Mathys J, Marchal K, Thijs G, De Moor B, Moreau Y (2002) Adaptive quality-based clustering of gene expression profiles. Bioinformatics 18:735–746

Diamanti-Kandarakis E, Bourguignon J-P, Giudice LC, Hauser R, Prins GS, Soto AM, Zoeller RT, Gore AC (2009) Endocrine-disrupting chemicals: an Endocrine Society scientific statement. Endocr Rev 30:293–342

Dickerson SM, Gore AC (2007) Estrogenic environmental endocrine-disrupting chemical effects on reproductive neuroendocrine function and dysfunction across the life cycle. Rev Endocr Metab Disord 8:143–159

Dickerson SM, Guevara E, Woller MJ, Gore AC (2009) Cell death mechanisms in GT1-7 GnRH cells exposed to polychlorinated biphenyls PCB74, PCB118, and PCB153. Toxicol Appl Pharmacol 237:237–245

Dickerson SM, Cunningham S, Patisaul HB, Woller MJ, Gore AC (2011) Endocrine disruption of brain sexual differentiation by developmental PCB exposure. Endocrinology 152:581–589

Dickerson SM, Cunningham SL, Gore AC (2011b) Prenatal PCBs disrupt early neuroendocine development of the rat hypothalamus. Toxicol Appl Pharmacol 252:36–46

Dumesic DA, Abbott DH, Padmanabhan V (2007) Polycystic ovary syndrome and its developmental origins. Rev Endocr Metab Disord 8:127–141

Esper RM, Pankonin MS, Loeb JA (2006) Neuregulins: versatile growth and differentiation factors in nervous system development and human disease. Brain Res Rev 51:161–175

Frame G (1997) A collaborative study of 209 PCB congeners and 6 Aroclors on 20 different HRGC columns, 2. Semi-quantitative Aroclor congener distributions. Fresenius J Anal Chem 357:714–722

Gore AC (2008) Developmental programming and endocrine disruptor effects on reproductive neuroendocrine systems. Front Neuroendocrinol 29:358–374

Gore AC (2010) Neuroendocrine targets of endocrine disruptors. Hormones 9:16–27

Gore AC, Crews D (2009) Environmental endocrine disruption of Brain and Behavior. In: Pfaff DW, Arnold AP, Etgen AM, Fahrbach SE, Rubin RT (eds) Hormones, Brain and Behavior, vol 3. Academic Press, San Diego, pp 1789–1816

Gore AC, Wu TJ, Oung T, Lee JB, Woller MJ (2002) A novel mechanism for endocrine-disrupting effects of polychlorinated biphenyls: direct effects on gonadotropin-releasing hormone neurones. J Neuroendocrinol 14:814–823

Hao P, He WZ, Huang Y, Ma LX, Xu Y, Xi H, Wang C, Liu BS, Wang JM, Li YX, Zhong Y (2005) MPSS: an integrated database system for surveying a set of proteins. Bioinformatics 21:2142–2143

Herbst AL, Ulfelder H, Poskanzer DC (1971) Adenocarcinoma of the vagina. Association of maternal stilbestrol therapy with tumor appearance in young women. N Engl J Med 284:878–881

Ho A, Sudhof TC (2004) Binding of F-spondin to amyloid-beta precursor protein: a candidate amyloid-beta precursor protein ligand that modulates amyloid-beta precursor protein cleavage. Proc Natl Acad Sci 101:2548–2553

Johansson M, Larsson C, Bergman A, Lund BO (1998a) Structure-activity relationship for inhibition of CYP11B1-dependent glucocorticoid synthesis in Y1 cells by aryl methyl sulfones. Pharmacol Toxicol 83:225–230

Johansson M, Nilsson S, Lund BO (1998b) Interactions between methylsulfonyl PCBs and the glucocorticoid receptor. Environ Health Perspect 106:769–772

Kerr MK, Martin M, Churchill GA (2000) Analysis of variance for gene expression microarray data. J Comput Biol 7:819–837

Kilic N, Sandal S, Colakoglu N, Kutlu S, Sevran A, Yilmaz B (2005) Endocrine disruptive effects of polychlorinated biphenyls on the thyroid gland in female rats. Tohoku J Exp Med 206:327–332

Landrigan PJ, Sonawane B, Butler RN, Trasande L, Callan R, Droller D (2005) Early environmental origins of neurodegenerative disease in later life. Environ Health Perspect 113:1230–1233

Lang IA, Galloway TS, Scarlett A, Henley WE, Depledge M, Wallace RB, Melzer D (2008) Association of urinary bisphenol A concentration with medical disorders and laboratory abnormalities in adults. JAMA 300:1303–1310

Layton AC, Sanseverino J, Gregory BW, Easter JP, Sayler GS, Schultz TW (2002) In vitro estrogen receptor binding of PCBs: measured activity and detection of hydroxylated metabolites in a recombinant yeast assay. Toxicol Appl Pharmacol 180:157–163

McLachlan JA, Newbold R, Shah HC, Hogan MD, Dixon RL (1982) Reduced fertility in female mice exposed transplacentally to diethylstilbestrol (DES). Fertil Steril 38:364–371

Meeks TW, Ropacki SA, Jeste DV (2006) The neurobiology of neuropsychiatric syndromes in dementia. Curr Opin Psychiatry 19:581–586

Miller DB, Gray LE Jr, Andrews JE, Luebke RW, Smialowicz RJ (1993) Repeated exposure to the polychlorinated biphenyl (Aroclor 1254) elevates the basal serum levels of corticosterone but does not affect the stress-induced rise. Toxicology 81:217–222

Petit F, Le Goff P, Cravédi JP, Valotaire Y, Pakdel F (1997) Two complementary bioassays for screening the estrogenic potency of xenobiotics: recombinant yeast for trout estrogen receptor and trout hepatocyte cultures. J Mol Endocrinol 19:321–335

Ptak A, Ludewig G, Lehmler HJ, Wojtowicz AK, Robertson LW, Gregoraszczuk EL (2005) Comparison of the actions of 4-chlorobiphenyl and its hydroxylated metabolites on estradiol secretion by ovarian follicles in primary cells in culture. Reprod Toxicol 20:57–64

Ptak A, Ludewig G, Robertson LW, Lehmler HJ, Gregoraszczuk EL (2006) In vitro exposure of porcine prepubertal follicles to 4-chlorobiphenyl (PCB3) and its hydroxylated metabolites: effects on sex hormone levels and aromatase activity. Toxicol Lett 164:113–122

Raisman G, Field PM (1971) Sexual dimorphism in the preoptic area of the rat. Science 173:731–733
Rehulka J, Minarik B (2004) Effect of polychlorinated biphenyls (Delor 103) on haematological and enzyme parameters of the rainbow trout Oncorhynchus mykiss. Dis Aquat Organ 62:147–153
Salama J, Chakraborty TR, Ng L, Gore AC (2003) Effects of polychlorinated biphenyls (PCBs) on female reproductive development and estrogen receptor β expression. Environ Health Perspect 111:1278–1282
Schrader TJ, Cooke GM (2003) Effects of Aroclors and individual PCB congeners on activation of the human androgen receptor in vitro. Reprod Toxicol 17:15–23
Scott CJ, Tilbrook AJ, Simmons DM, Rawson JA, Chu S, Fuller PJ, Ing NH, Clarke IJ (2000) The distribution of cells containing estrogen receptor α (ERα) and ERβ messenger ribonucleic acid in the preoptic area and hypothalamus of the sheep: comparison of males and females. Endocrinology 141:2951–2962
Seegal RF (1996) Epidemiological and laboratory evidence of PCB-induced neurotoxicity. Crit Rev Toxicol 26:709–737
Sheehan DM (2006) No threshold dose–response curves for nongenotoxic chemicals: findings and applications for risk assessment. Environ Res 100:93–99
Shekhar PV, Werdell J, Basrur VS (1997) Environmental estrogen stimulation of growth and estrogen receptor function in preneoplastic and cancerous human breast cell lines. J Natl Cancer Inst 89:1774–1782
Smith JT, Cunningham MJ, Rissman EF, Clifton DK, Steiner RA (2005) Regulation of Kiss1 gene expression in the brain of the female mouse. Endocrinology 146:3686–3692
Steinberg RM, Juenger TE, Gore AC (2007) The effects of prenatal PCBs on adult female paced mating reproductive behaviors in rats. Horm Behav 51:364–372
Steinberg RM, Walker DM, Juenger TE, Woller MJ, Gore AC (2008) Effects of perinatal polychlorinated biphenyls on adult female rat reproduction: development, reproductive physiology, and second generational effects. Biol Reprod 78:1091–1101
Tetel MJ (2009) Nuclear receptor coactivators: essential players for steroid hormone action in the brain and in behaviour. J Neuroendocrinol 21:229–237
Tobet SA, Hanna IK (1997) Ontogeny of sex differences in the mammalian hypothalamus and preoptic area. Cell Mol Neurobiol 17:565–601
Troster AI, Ruff RM, Watson DP (1991) Dementia as a neuropsychological consequence of chronic occupational exposure to polychlorinated biphenyls (PCBs). Arch Clin Neuropsychol 6:301–318
Walker DM, Juenger TE, Gore AC (2009) Developmental profiles of neuroendocrine gene expression in the preoptic area of male rats. Endocrinology 150:2308–2316
Wang XQ, Fang J, Nunez AA, Clemens LG (2002) Developmental exposure to polychlorinated biphenyls affects sexual behavior of rats. Physiol Behav 75:689–696
Weltje L, vom Saal FS, Oehlmann J (2005) Reproductive stimulation by low doses of xenoestrogens contrasts with the view of hormesis as an adaptive response. Hum Exp Toxicol 24:431–437
Winneke G, Bucholski A, Heinzow B, Krämer U, Schmidt E, Walowiak J, Wiener J-A, Steingrüber H-J (1998) Developmental neurotoxicity of polychlorinated biophenyls (PCBS): cognitive and psychomotor functions in 7-month old children. Toxicol Lett 102–103:423–428
Woodhouse AJ, Cooke GM (2004) Suppression of aromatase activity in vitro by PCBs 28 and 105 and Aroclor 1221. Toxicol Lett 152:91–100

The Kisspeptin System as Putative Target for Endocrine Disruption of Puberty and Reproductive Health

Manuel Tena-Sempere

Abstract The activation of the reproductive axis at puberty and its proper function later on life are founded on a complex series of maturational events that include the sexual differentiation of the brain during early *critical* periods of development. Brain sex differentiation is driven (mostly) by endogenous sex steroids, which are also important regulators of the neuroendocrine networks governing puberty onset. Accordingly, both phenomena might be sensitive to the disrupting actions of exogenous compounds with sex steroid-like activity that may pose long lasting consequences in terms of reproductive health. Kisspeptins, the products of the *Kiss1* gene that act via the receptor, GPR54, are neuropeptides produced at discrete neuronal populations within the hypothalamus with key roles in brain sex differentiation, puberty onset and fertility. A subset of Kiss1 neurons has been recently shown to co-express neurokinin B (NKB), another neuropeptide with important reproductive roles. Compelling evidence has mounted recently that Kiss1 neurons are subjected to sexual dimorphism and physiologically sensitive to the *organizing* and *activational* effects of sex steroids, as documented in rodents, sheep and primates (including humans). These features provide the basis for the potential endocrine disruption of reproductive maturation and function by xeno-steroids targeting the Kiss1 system. Indeed, solid, as yet fragmentary, evidence obtained in rodents and sheep suggests that hypothalamic expression of *Kiss1* and/or kisspeptin fiber distribution are altered following inappropriate exposures to synthetic estrogenic and/or androgenic compounds during critical periods of development. Of note, administration of androgenic and estrogenic compounds during such critical periods has been shown to alter also the hypothalamic expression of NKB in rodents and sheep. As a whole, the data summarized in this chapter document the sensitivity of Kiss1 system to changes in sex

M. Tena-Sempere (✉)
Department of Cell Biology, Physiology and Immunology, University of Córdoba; CIBERobn Fisiopatología de la Obesidad y Nutrición; and Instituto Maimonides de Investigaciones Biomédicas de Córdoba (IMIBIC) and Physiology Section, Córdoba, Spain
e-mail: fi1tesem@uco.es

J.-P. Bourguignon et al. (eds.), *Multi-System Endocrine Disruption*, Research and Perspectives in Endocrine Interactions,
DOI 10.1007/978-3-642-22775-2_2, © Springer-Verlag Berlin Heidelberg 2011

steroid milieu during critical periods of sexual maturation, and strongly suggest that alterations of endogenous kisspeptin tone induced by inappropriate early exposures to environmental compounds with sex steroid activity might be mechanistically relevant for disruption of puberty onset and gonadotropin secretion later in life. The potential additional interactions of xeno-hormones with related neuropeptide systems (including NKB) and other environmental modulators (such as the nutritional state) of the Kiss1 system merits future investigation.

Introduction

During the last two decades, considerable attention has focused on the potential adverse effects for humans and wild-life species of a variety of natural and synthetic compounds, with ability to mimic or interfere the biological activities of endogenous hormones, thereby *disrupting* the development and/or function of different endocrine axes (Toppari et al. 1996). These are globally named endocrine disrupting compounds (EDCs), and include different man-made chemicals and by-products, as well as plant compounds, with either estrogenic, androgenic or anti-androgenic activity (Kelce and Gray 1999; Sikka and Naz 1999; Tena-Sempere et al. 2000). By far, the greatest concerns for the potential adverse effects of EDCs have concentrated on their deleterious impact on reproductive health. Indeed, different epidemiological studies have documented the recent worsening of several reproductive parameters, ranging from decreased sperm counts to increased incidence of testicular cancer and genital malformations, in different species, including (prominently) humans (Guillette 2006; Norgil Damgaard et al. 2002; Skakkebaek et al. 2006). Although multifaceted, this phenomenon is likely to involve environmental factors, including increasing exposures to different EDCs.

Among the potential reproductive end-points of endocrine disruption, the direct adverse effects of EDCs on the developing gonads and genitalia have been predominantly scrutinized, mostly because of the clinical manifestations (decreased sperm production; increased frequency of gonadal cancer and genital malformations) indicated above. More recently, however, experimental evidence has begun to accumulate suggesting that some suspected forms of endocrine disruption in humans and other species, such as changes in the tempo of puberty and some forms of infertility, may originate from primary defects not in the gonads and/or genitalia but rather in the development and/or function of the central (hypothalamic) systems controlling the reproductive axis (Bourguignon et al. 2010; Diamanti-Kandarakis et al. 2009; Gore 2008a). In this chapter, we will critically review the compelling, although as yet scarce, experimental data suggesting that the Kiss1 system might be a prominent target for disruption of reproductive health by inappropriate exposures to EDCs with sex-steroid like activity. As a means to provide a mechanistic basis for such a possibility, we will first summarize briefly our current knowledge on the essential roles of this hypothalamic system in the control of key aspects of reproductive maturation and function, and will recapitulate the available experimental evidence supporting its physiological modulation by sex steroids along the lifespan.

Endocrine Disruption of Central Elements of the Reproductive Axis: Lessons from Physiology

Attainment of reproductive capacity at puberty, and its maintenance during adulthood, is the final consequence of a complex series of maturation events that involve proper functional organization at early stages of development of the central (hypothalamic) circuitries responsible for the control of the pulsatile secretion of gonadotropin-releasing hormone (GnRH), the major driving signal of pituitary gonadotropin synthesis and release (Fink 2000; Gore 2008a; Tena-Sempere and Huhtaniemi 2003). A first key component of this developmental phenomenon is the process of sexual differentiation of the reproductive brain; an event which is closely related with, but genuinely different from, other sex-determination phenomena, such a gonadal and genital differentiation (Gore 2008a). An important feature of brain sex differentiation is that it is sexually dimorphic and mainly driven by sex steroids. Thus, in rodents, where the molecular basis of this phenomenon have been deeply scrutinized, it is known that the process of brain sex differentiation takes place during a critical developmental period from late embryonic to early postnatal age (Tena-Sempere et al. 2000), when key neuronal networks at the hypothalamus become *organized* in a permanent manner differentially in males and females (Gore 2008a; Morris et al. 2004).

Rodent data have firmly demonstrated that the major molecular signal responsible for brain *masculinization* is estradiol, locally produced at high levels by aromatization of testis-derived testosterone (Gore 2008a; MacLusky and Naftolin 1981; Morris et al. 2004). In the female, low estrogen input due to quiescent ovaries and high levels of circulating α-fetoprotein is the major determinant of brain *feminization* (Gore 2008a). Importantly, such early differentiation phenomena translate into relevant sexually-dimorphic features, such as the timing of puberty (Ojeda et al. 2006, 2010), as well as different behaviors and neuroendocrine secretory patterns later on life (Gore 2008a; Morris et al. 2004). A hallmark of such neurohormonal sexual dimorphism is the cyclic secretory activity of the GnRH/gonadotropin system, which is based on the capacity of estrogens, selectively in the female, to induce the pre-ovulatory surge of gonadotropins from puberty onwards (Christian and Moenter 2010; Gore 2008a).

Recent experimental evidence suggests that, in addition to earlier (perinatal) periods, puberty itself may represent a second critical window for neuroendocrine development, during which changes in sex steroid input might induce permanent functional alterations of different neuro-hormonal axes later in life (Evuarherhe et al. 2009). Indeed, important developmental effects of estrogen on key hypothalamic networks governing puberty onset have been recently documented during (pre)pubertal maturation in mice (see Sect. "The Kiss1 System: Essential Regulator of Brain Sex Differentiation and Puberty"). In addition to such potential organizing effects, it is well known that the pubertal transition is characterized by substantial acute changes in the sensitivity and responsiveness of reproductive hypothalamic networks to the regulatory effects of sex steroids (Ebling 2005). These changes importantly contribute, together with gonadal-independent variations of the central excitatory and inhibitory inputs to GnRH neurons (Ojeda et al. 2006, 2010), to define the timing of puberty.

Considering the above physiological features, it is plausible that xeno-hormones with sex steroid-like activities may impact at central levels to interfere with proper sexual differentiation and/or later maturation of the reproductive axis (Bourguignon et al. 2010; Diamanti-Kandarakis et al. 2009; Gore 2008a, b; Navarro and Tena-Sempere 2008). Yet, despite recent progress in this area, the mechanisms and targets for such a potential disruption of the reproductive neuroendocrine systems remain incompletely characterized. In this context, the use of experimental animal models, mimicking inappropriate early exposures to compounds with sex steroid activity, has provided valuable information (Tena-Sempere et al. 2000), and may prove helpful for further dissection of the central effects and mechanisms of action of specific EDCs. Indeed, these models may represent a good complement to robust *in vitro* and *ex vivo* approaches, commonly used for the evaluation of single EDC effects on key components of the reproductive brain, such as GnRH neurons themselves (Bourguignon et al. 2010; Gore 2008a). Indeed, neuroendocrine analyses in rodents subjected to protocols of neonatal exposure to estrogenic compounds, such as estradiol benzoate (EB) and diethylstilbestrol (DES), have revealed that disruption of sexual differentiation of the hypothalamus is linked to altered puberty onset and disturbed gonadotropin secretion, both at basal conditions and after gonadectomy (Pinilla et al. 1995; Tena-Sempere et al. 2000). Likewise, protocols of early postnatal exposure of female rats to estradiol and the estrogenic EDC, dichlorodiphenyltrichloroethane (DDT), have documented alterations in hypothalamic GnRH secretion by hypothalamic explants *ex vivo* and LH responses to GnRH *in vivo* in infantile/juvenile rats (Rasier et al. 2007). These studies, among others [for an extensive review see (Bourguignon et al. 2010; Diamanti-Kandarakis et al. 2009; Gore 2008a), and references therein], set the scene for specific mechanistic analyses on the impact and mode of action of 'actual' EDCs on the hypothalamic systems controlling reproduction, some of which have been recently initiated (Bourguignon et al. 2010; Gore 2008a). The potential involvement of the hypothalamic Kiss1 system in this phenomenon is proposed, on the basis of the available literature, in the following sections.

The Kiss1 System: Essential Regulator of Brain Sex Differentiation and Puberty

The so-called Kiss1 system has been recently recognized as an essential regulator of the reproductive axis, which is indispensable for its timely activation at puberty and its proper function (and hence, fertility) during adult life (Oakley et al. 2009;

Roa et al. 2008). Remarkably, the elements of this ligand-receptor system were sequentially identified between late 1990s and early 2000s in a field totally unrelated with reproduction. Thus, the *Kiss1* gene was originally catalogued as metastasis suppressor, as it was found to encode a number of structurally-related peptides, globally termed kisspeptins, with ability to suppress tumor spread (Oakley et al. 2009; Roa et al. 2008). Already in 2001, kisspeptins were shown to conduct their biological actions via the G protein-coupled receptor, GPR54 (Kotani et al. 2001; Muir et al. 2001; Ohtaki et al. 2001). However, recognition of the reproductive dimension of this system only came in late 2003, when inactivating mutations of GPR54 were found in patients suffering idiopathic forms of hypogonadotropic hypogonadism (de Roux et al. 2003; Seminara et al. 2003); findings that were replicated and extended in mice bearing null mutations of *GPR54* or *Kiss1* genes (d'Anglemont de Tassigny et al. 2007; Seminara et al. 2003). These seminal observations boosted an enormous interest in the area and initiated an ever growing number of studies aiming to characterize the physiological roles and mechanisms of action of kisspeptins and GPR54 in the control of different facets of reproductive function. As a whole, these studies have firmly substantiated the pivotal functions of kisspeptins in the control of the reproductive brain in different mammalian and non-mammalian species (d'Anglemont de Tassigny and Colledge 2010; Oakley et al. 2009; Popa et al. 2008; Roa et al. 2008). While the rapid expansion of this area makes it impossible to comprehensively condensate the state-of-the-art in this particular field within the limits of this chapter, we provide in this section some brief description on the brain distribution and major mechanisms of action of kisspeptins, together with a summary of their putative roles in sex differentiation and puberty onset, as a means to introduce and to ease the comprehension of later sections of this review.

Recent neuroanatomical studies, carried out mostly in laboratory rodents, but also in sheep and primates (including humans), have identified discernible populations of Kiss1 neurons, and some of their projections, at the hypothalamus (Herbison 2008; Popa et al. 2008; Roa et al. 2008). In the rat and mouse, where extensive mapping of these neuronal populations has been published, hypothalamic Kiss1 neurons appear to concentrate mostly at two areas: the arcuate nucleus (ARC) and the anteroventral periventricular nucleus (AVPV) (Popa et al. 2008); the latter population appears to extend as a continuum along the rostral peri-ventricular area of the third ventricle (RP3V) (Herbison 2008). Prominent groups of Kiss1 neurons have been also detected in the ARC (or the equivalent infundibular area) in sheep and primates (Oakley et al. 2009; Roa et al. 2008); species where the presence and eventual physiological roles of a rostral population of Kiss1 neurons is still under some debate. Similarly, while direct synaptic contacts have been described between Kiss1 and GnRH neurons (Clarkson and Herbison 2006), the neuroanatomical features of the projections of Kiss1 neurons originating from the ARC and the AVPV need further clarification.

A wealth of experimental evidence, accumulated during the last 5 years, has demonstrated that the primary mechanism whereby kisspeptins participate in the control of the reproductive axis involves its ability to operate upon, and stimulate, GnRH neurons, which have been shown to express GPR54 (d'Anglemont de Tassigny and Colledge 2010; Oakley et al. 2009; Popa et al. 2008; Roa et al. 2008). In fact, the potent stimulatory effects of kisspeptins on luteinizing hormone (LH) and follicle-stimulating hormone (FSH) secretion are blocked by pre-treatment with a potent GnRH antagonist, while kisspeptin administration is able to activate (as reflected by c-fos induction) GnRH neurons. Moreover, kisspeptins have been reported to induce very potent depolarization responses in GnRH neurons, as well as the release of GnRH by hypothalamic preparations *ex vivo* and to the Cerebrospinal Fluid (CSF) *in vivo* (d'Anglemont de Tassigny and Colledge 2010; Oakley et al. 2009; Popa et al. 2008; Roa et al. 2008). Admittedly, whether these stimulatory effects take place preferentially at GnRH neuronal perikarya (in the preoptic area) and/or nerve terminals (in the median eminence) is yet to be defined. Regardless of the actual site(s) of action on GnRH neurons, the potent stimulatory effects of kisspeptins on GnRH secretion result in robust LH and FSH responses, as documented in different mammalian species, even at very low doses (Oakley et al. 2009; Roa et al. 2008).

Indirect evidence for the potential involvement of the populations of Kiss1 neurons in the process of brain sex differentiation came from comparative neuroanatomical studies in male and female rodents and primates. Thus, *Kiss1* mRNA expression has been shown sexually dimorphic at the AVPV in rats, with females having higher expression than males (Kauffman et al. 2007). A similar dimorphism appears to exist also in humans, where the populations of Kiss1 neurons at the infundibular and periventricular regions are more abundant in the female (Hrabovszky et al. 2010). Whether Kiss1 neurons at the ARC are also sexually dimorphic in number in rodents is presently under investigation; although initial reports documented that *Kiss1* mRNA levels were similar between sexes, recent immunohistochemical studies suggested that kisspeptin content at this hypothalamic site is probably higher in female rats also. As a whole, the above data evidence that the populations of Kiss1 neurons undergo a program of sexual differentiation, likely during early stages of development, which enables sex-specific configurations of Kiss1 networks later in life. Such a neuronal set-up appears to be functionally relevant for the expression of sexually-dimorphic neuroendocrine traits later in life, such as the positive feedback of estrogen and the pre-ovulatory surge of gonadotropins. The importance of kisspeptin signaling in this phenomenon is reinforced by the fact that pharmacological blockade of kisspeptin actions results in elimination of the pre-ovulatory peak of gonadotropins in cyclic female rats (Pineda et al. 2010).

Likewise, kisspeptins play an essential role in the regulation of timing of puberty, as evidenced by a combination of genetic, physiologic and pharmacological analyses in rodents and primates (Tena-Sempere 2010b). These analyses have documented that the hypothalamic Kiss1 system undergoes a complex and sophisticated activation program during puberty, which seems to involve four major components: (a) the elevation in the endogenous kisspeptin tone at the hypothalamus (Navarro et al. 2004; Shahab et al. 2005), (b) the increase in the sensitivity to the stimulatory effects of kisspeptins in terms of GnRH/LH responses (Castellano et al. 2006; Han et al. 2005), (c) the enhancement of GPR54 expression and signaling efficiency (Han et al. 2005; Herbison et al. 2010), and (d) the elevation of the number of kisspeptin neurons and of their projections to GnRH neurons (Clarkson et al. 2009; Clarkson and Herbison 2006). The physiological relevance of the above maturational changes for the proper timing of puberty is stressed by the observation that selective blockade of kisspeptin actions during the pubertal transition significantly delays puberty onset in female rats (Pineda et al. 2010).

The Kiss1 System: Sensitivity to the *Activational* and *Organizing* Effects of Sex Steroids

Shortly after the disclosure of its reproductive dimension, sex steroids (mainly, estradiol and testosterone) were recognized as important regulators of hypothalamic *Kiss1* gene expression during adulthood. This important regulatory role was first evidenced by experiments of gonadectomy, with or without sex steroid replacement, in rodents, and has been later confirmed in different mammalian species, including primates (Oakley et al. 2009; Roa et al. 2008). A similar regulatory action is operative also in humans, as demonstrated by studies on the changes in hypothalamic *Kiss1* mRNA levels in postmenopausal women (Rance 2009). While detailed description of this phenomenon exceeds the limits of this review, it is important to stress two major features related with the activational regulation of hypothalamic *Kiss1* expression by sex steroids: (a) their effects are, at least in rodents, clearly nucleus-specific, since sex steroids suppress *Kiss1* mRNA levels at the ARC, but estrogen enhances *Kiss1* expression at the AVPV; and (b) given the above differential regulation, the ARC Kiss1 pathway has been involved in the negative feedback control of gonadotropin secretion, whereas Kiss1 neurons at the AVPV may mediate the positive feedback effects of estrogen, which are responsible for the pre-ovulatory surge of gonadotropins (Oakley et al. 2009; Roa et al. 2008).

In addition to the above activational effects, compelling evidence has mounted recently that the developmental changes of the Kiss1 system, summarized in the previous section, are shaped to a large extent by the organizing and regulatory actions of sex steroids acting during critical periods of perinatal or postnatal development. This contention has been documented for the process of brain sex differentiation in rodents. Indeed, studies in rats and mice have substantiated that changes in sex steroid inputs to the developing brain during early critical periods of maturation alter the normal process of sexual differentiation of Kiss1 neuronal populations. Thus, neonatal exposure to high doses of androgen in female rats suppressed the expression of *Kiss1* mRNA at the AVPV in adulthood, when neonatally androgenized females displayed *Kiss1* mRNA levels similar to those of adult males, but much lower than in cyclic females (Kauffman et al. 2007). Conversely, neonatal orchidectomy in rats, which eliminated androgen secretion (and hence, actions) during the critical period of brain sex differentiation, resulted in the feminization of *Kiss1*/kisspeptin expression at the AVPV, which was much higher than in control adult males (Homma et al. 2009). The functional relevance of the above changes is stressed by the observation that neonatally androgenized females were devoid of positive feedback actions of estrogen in adulthood (Kauffman et al. 2007), while neonatal orchidectomy resulted in the acquisition of positive feedback and surge-like LH responses to ovulatory doses of estradiol (Homma et al. 2009).

In addition to androgens, estrogen has been also documented as an important organizing signal for the hypothalamic Kiss1 system, acting during critical windows of development. Thus, neonatal estrogenization of female rats decreased the number of Kiss1 neurons at the AVPV and disrupted the positive feedback effects of estradiol (in terms of LH secretion) in adulthood (Homma et al. 2009). Similarly, neonatal exposures to synthetic estrogens, known to disturb proper activation and function of the gonadotropic axis (Tena-Sempere et al. 2000), persistently suppressed the hypothalamic expression of *Kiss1* gene at the expected time of puberty and adulthood in male and female rats (Navarro et al. 2004, 2009). In good agreement, studies in α-fetoprotein (AFP) knock-out mice, which are developmentally exposed to excessive estrogenic input due to the lack of this scavenger protein of circulating estrogens, have demonstrated that in the female such an excessive exposure to estrogen results in lower numbers of Kiss1 neurons and perturbs the ability of estradiol to induce pre-ovulatory-like LH surges and to activate Kiss1 neurons at the AVPV in adulthood (Gonzalez-Martinez et al. 2008). Admittedly, however, mechanistic interpretation of the above results is potentially confounded by the fact that this KO model does not allow proper discrimination between early organizing versus later activational actions of sex steroids.

Besides its organizing effects during early (perinatal) development, estrogen appears to be important also for shaping the expansion of the AVPV population of Kiss1 neurons in the female mouse during the pubertal transition, as documented by the effects of ovariectomy during the late infantile period (Clarkson et al. 2009). Likewise, in aromatase null (ArKO) mice, where transformation of androgen into estrogen is blocked, the pubertal increase of Kiss1 neurons at the AVPV was prevented (Clarkson et al. 2009), and kisspeptin-IR at this site was substantially reduced at adulthood (Bakker et al. 2010). As a whole, these data would imply that some sort of positive feedback of estrogen on the hypothalamic Kiss1 system may exist before puberty. It remains to be defined which level of estrogenic input is required for the (proper) pubertal activation of the Kiss1 system. Similarly, whether inappropriate exposures to sex steroid-acting compounds during the pubertal transition have long-lasting consequences in terms of configuration, and eventual function, of the populations of Kiss1 neurons at the hypothalamus awaits further investigation.

The Kiss1 System as Target of Endocrine Disruption: Experimental Evidence

As reviewed in previous sections, a large body of experimental data has accumulated in recent years to demonstrate that Kiss1 neurons at specific hypothalamic nuclei are putative components of the mechanisms of brain sex differentiation, and highly sensitive to the organizing actions of endogenous, and eventually exogenous, sex steroids. This phenomenon, which is of substantial physiological relevance, may pose also potential pathophysiological implications, as it provides the neurohormonal substrate for the perturbation of the development and function of the gonadotropic axis by the impact of EDCs on the developing Kiss1 system, acting at early, critical periods of maturation (Tena-Sempere 2010a). Admittedly, however, the available evidence is still fragmentary and mostly derived from analyses of standard models of perinatal estrogenization or androgenization in rodents, rather than '*real-life*' protocols of exposure to EDCs. Notwithstanding, considering the recommendation to enhance basic, mechanistic studies and to apply the precautionary principle in the field of endocrine disruption research (Diamanti-Kandarakis et al. 2009), a succinct critical review of the available evidence is provided below, as a means to pave the way for further analyses in this area.

In the rat, protocols of neonatal exposure to the synthetic estrogen, estradiol benzoate (EB), have been shown to induce a persistent decrease in hypothalamic expression of *Kiss1* gene at the time preceding puberty in male and female rats (Navarro et al. 2009). This phenomenon, which was also observed in adult animals submitted to neonatal exposure to EB (Navarro et al. 2004), was dose-dependent, with detectable effects in terms of suppressed *Kiss1* mRNA levels at a range of doses (1–10 μg; on d-1 postpartum) that is considered moderate as compared to standard protocols of neonatal estrogenization in rodents. Of note, hypothalamic *Kiss1* responses to gonadectomy (i.e., increase in its mRNA levels) were also blunted in neonatally estrogenized rats, suggesting that disturbed responses to negative gonadal feedback following early exposures to estrogenic compounds might involve also altered expression and/or function of the Kiss1 system (Navarro et al. 2009). In good agreement, neonatal exposures to EB resulted in significantly decreased kisspeptin fiber density in the ARC and AVPV in female rats at adulthood; effects that were partially mimicked by neonatal treatments of females with the selective ligand of estrogen receptor (ER)α, PPT, that decreased kisspeptin fibers at the ARC, and the phytoestrogen, genistein, which suppressed kisspeptin fiber density at AVPV in adulthood (Bateman and Patisaul 2008). Indeed, a recent study has demonstrated that neonatal exposure to a high dose of genistein, as it is the case also for EB, disturbs the normal pattern of postnatal expansion of kisspeptin fiber density at the AVPV (and the ARC) in female rats; neonatally treated rats having persistently lower numbers of kisspeptin fibers than controls along the pubertal transition (Losa et al. 2011).

In the same vein, neonatal exposure to the xeno-estrogen, bisphenol A (BPA), in rats significantly reduced hypothalamic *Kiss1* mRNA levels at puberty (Navarro et al. 2009), as well as kisspeptin immunoreactivity at the ARC (as estimated in terms of fiber density) in adult female rats (Patisaul et al. 2009). Admittedly, however, the effects of BPA were detected at rather high doses and were not observed in males, which appeared less sensitive to the potential disrupting effects of this EDC (Patisaul et al. 2009). Also of note, recent, as yet preliminary, evidence obtained in mouse off-spring from mothers exposed orally to BPA suggest that, in some conditions, this xeno-estrogen might induce also up-regulatory actions on kisspeptin content at the hypothalamus. Indeed, such protocol of exposure to BPA has been reported to enhance kisspeptin expression of kisspeptin at the ARC and AVPV, especially in male mice; a response that partially obliterated the sexual dimorphism (females >> males) that is usually observed, in terms kisspeptin content, at the hypothalamus of adult rodents (Panzica et al. 2009).

In any event, it is important to stress that the impact of potential EDCs on the hypothalamic Kiss1 system has not been only documented in rodents but also in sheep. Thus, recent data demonstrated that in utero exposure of sheep to a complex cocktail of EDCs contained in sewage sludge, used as agricultural fertilizer in pastures and thus an optimal model for investigation of 'real-life' mixtures of ED, induced a significant decrease in *Kiss1* mRNA levels at the rostral, mid and caudal regions of the hypothalamus of exposed fetuses (Bellingham et al. 2009). Remarkably, no effect was observed when exposures were conducted during adulthood, thus emphasizing the notion that the Kiss1 system may be especially vulnerable to the disrupting actions of compounds with sex steroid activities when exposures take place during critical periods of maturation (Bellingham et al. 2009). As additional note, it is to be stressed that the above protocols of fetal sewage sludge exposures resulted also in reduced numbers of kisspeptin immunopositive cells at the pituitary (Bellingham et al. 2009). In this sense, expression and (activational) sex steroid regulation of Kiss1 and GPR54 mRNAs/proteins have been previously demonstrated in the rat pituitary (Richard et al. 2008). This observation raises the appealing possibility of additional sites, other than the hypothalamus, for the disrupting neuroendocrine effects of EDCs on the Kiss1 system. Admittedly, however, the physiological relevance of kisspeptin actions and regulation directly at the pituitary level is yet to be fully unfolded (Roa et al. 2008).

All in all, the data summarize above provide strong, as yet circumstantial, evidence for the possibility that inappropriate exposures to sex-steroid acting compounds during early critical windows of brain sex differentiation might have a *disorganizing* impact on the maturation of the hypothalamic Kiss1 system, with durable consequences that may manifest later in life (i.e., during puberty and adulthood). In keeping with such possibility, functional studies conducted in some of the above models have conclusively shown that the decreased pubertal expression of *Kiss1* gene following neonatal exposures to estrogenic compounds is associated to defective gonadotropin secretion (both in basal and post-gonadectomy conditions), which can be rescued by administration of exogenous kisspeptin (Navarro et al. 2009). Similarly, some of the above protocols of neonatal exposure to chemicals with estrogenic actions linked to decreased kisspeptin fiber density at certain hypothalamic nuclei in adult female rats caused also reduced GnRH neuronal activation (Bateman and Patisaul 2008). Admittedly, however, defective activation of GnRH neurons has not been detected following neonatal exposures to other estrogenic compounds, such as BPA (Adewale et al. 2009). In any event, these functional data strongly suggest that by targeting the developing hypothalamic Kiss1 system, and thereby by presumably causing a decrease in the endogenous kisspeptin tone, compounds with sex steroid-like activities might interfere with normal maturation and later function of the reproductive axis. While the evidence supporting this possibility is compelling, it must be emphasized, as general call of caution, that most of the work conducted so far in this area has addressed putative mechanisms of disruption rather than the consequences of '*real-life*' exposures to complex mixtures of ED, presumably at low doses (Tena-Sempere 2010a).

Other Central (Related) Targets for Endocrine Disruption: A Case for NKB?

Although abundant physiologic evidence and growing pharmacological data support the contention that the hypothalamic Kiss1 system might be a potential target for endocrine disruption, there are strong experimental indications that EDCs may have an impact also upon other key elements of the reproductive brain, whereby they may contribute to alter basic neuroendocrine networks responsible for the proper maturation and function of the reproductive axis (Bourguignon et al. 2010; Diamanti-Kandarakis et al. 2009; Gore 2008a). While thorough recapitulation of such evidence is beyond the scope and aims of this review, it is important to stress that a wealth of data has documented the potential impact of different EDCs, such as dichlorodiphenyltrichloroethane (DDT), vinclozolin and certain polychlorinated biphenils (PCBs), as well as complex mixtures of EDCs (present in sewage sludge) on the development, neurosecretory function and even viability/apoptosis of the GnRH system in different species (Bellingham et al. 2010; Bourguignon et al. 2010; Dickerson et al. 2009; Rasier et al. 2007; Wadas et al. 2010), as assessed by a combination of *in vitro*, *in vivo* and *ex vivo* approaches. Of note, the effects of EDCs, such as DDT, on GnRH secretory function appears to be multifaceted, and involve the modulation of glutamate stimulation of GnRH neurons via multiple pathways (Bourguignon et al. 2010). In addition to GnRH neurons themselves, fragmentary evidence has suggested the impact of early exposures to putative EDCs on other hypothalamic pathways and signals, such as galanin, in sheep fetuses exposed to sewage sludge (Bellingham et al. 2010), and ocytocin, in female rats exposed neonatally to BPA (Adewale et al. 2011). As a whole, the above data illustrate the potential complexity of the neuroendocrine effects of EDCs, which may impact at different levels/elements of the reproductive brain. Elucidation of the complete set of targets for such compounds appears critical to provide an integral mechanistic insight into their mode of action and potential adverse effects upon the reproductive axis.

Very recently, NKB has emerged as co-transmitter and putative auto-regulator of the population of Kiss1 neurons located at the ARC/infundibular region in different species, including rodents, sheep and humans (Lehman et al. 2010). The functional relevance of NKB in reproductive control has gained momentum with the observation that humans with inactivating mutations in the genes encoding NKB (*TAC3*) or its receptor (*TACR3*) suffer hypogonadotropic hypogonadism (Topaloglu et al. 2009); hence, a phenotype similar to that of GPR54 null patients. In keeping with such a *positive* role in the control of the reproductive axis, very recent experimental data have documented the stimulatory effects of agonists of NKB on LH secretion in female rats, sheep and monkeys (Billings et al. 2010; Navarro et al. 2011; Ramaswamy et al. 2010); yet, lack of stimulatory effects or even inhibitory actions of NKB on gonadotropin secretion have been also reported in some species and physiological conditions. The demonstration that Kiss1 neurons at the ARC co-express and are targets of NKB (Lehman et al. 2010), together with its putative effects on the GnRH pulse generator (Wakabayashi et al. 2010), has led to the contention that NKB participates in the fine tuning of kisspeptin output at the hypothalamus, thereby playing a role in the regulation of pulsatile GnRH secretion. Given the close relationship with the Kiss1 system, the question arises as whether NKB might be an additional target for endocrine disruption of the reproductive brain.

Admittedly, the above issue has not been directly addressed since, to our knowledge, the impact of early exposures to EDCs on the developmental expression of NKB has not been reported to date. Yet, the physiological knowledge available strongly supports such a possibility. First, hypothalamic expression of NKB (mRNA/protein) at the ARC is under the regulation of sex steroids in adulthood, as documented in models of gonadectomy, with a clear parallelism between Kiss1 and NKB expression. Thus, in female rats, estrogen suppressed *NKB* (as well as *Kiss1*) mRNA expression, therefore suggesting a role in negative feedback control (Navarro et al. 2010). Even more interestingly from the perspective of endocrine disruption, NKB expression in the ARC appears to be sensitive to the organizing effects of sex steroids, as documented in the sheep and rodents. Thus, in the sheep, the expression of NKB is sexually dimorphic (with females having double the number of NKB neurons than males) and was markedly suppressed in the female by prenatal exposure to testosterone (Cheng et al. 2010). Similarly, in the rat, NKB neurons at the ARC are responsive to sex steroids at different developmental stages (including puberty), with an androgen-dependent sexual dimorphism in the postnatal ontogeny of NKB peptide expression, that increased earlier in females along pre-pubertal maturation (Ciofi et al. 2007). Similarly, our preliminary evidence suggests the neonatal estrogenization suppresses both kisspeptin and NKB immunoreactivity at the ARC in adulthood more prominently in females, and that LH responses to the NKB agonist, senktide, are sexually dimorphic in the rat, with stimulatory responses being detected only in adult females, but not in adult male rats (*our unpublished data*). In this context, the possibility that developmental exposures to sex steroid-acting compounds may have a durable impact on the organization of the hypothalamic NKB system is plausible and merits specific investigation. Noteworthy, protocols of prenatal androgenization in the female sheep have been reported to suppress NKB expression, but not kisspeptin immunoreactivity, at the ARC (Cheng et al. 2010). Therefore, it is possible that, depending on the species, type of compound and developmental window of exposure, sex steroids compounds might variably affect the maturation and/or expression of Kiss1 and NKB systems, as putative mechanism for alteration of the reproductive brain.

Open Questions and Future Directions

As reviewed in previous sections, accumulating (although still limited) experimental evidence points out that the hypothalamic Kiss1 system may be a target for endocrine disrupting compounds with activity as estrogens or anti-androgens, acting at critical windows of development. While the physiological features of this system, as exquisitely responsive to sex steroids at different developmental stages and fundamental for proper puberty onset and fertility, make this possibility highly plausible, it must be stressed that most of the work conducted so far in this area has addressed putative mechanisms of disruption (i.e., if this phenomenon is biologically possible) rather than the consequences of 'real-life' exposures (i.e., whether this phenomenon is actually taking place). In other words, while it is tenable that EDCs may target the developing Kiss1 system, the pathophysiological relevance of such phenomenon in terms of reproductive health warrants additional experimental work, involving exposures to complex cocktails of EDC with similar or dissimilar modes of action, ideally at low doses, in line with very recent studies initiated in the sheep (Bellingham et al. 2009). Admittedly, most of the studies on the neuro-endocrine effects of EDCs have focused on their impact on different aspects of GnRH neuronal development and secretory function (Bourguignon et al. 2010; Diamanti-Kandarakis et al. 2009; Gore 2008a). In any event, identification of the potential impact of such compounds upon Kiss1 neurons may allow to revisit and to provide a mechanistic insight for previously documented neuroendocrine effects of some EDCs, which were reported prior to the recognition of the prominent reproductive roles of the Kiss1 system. For instance, it was described that neonatal exposures to genistein and BPA alter sexual differentiation of AVPV in rats (Patisaul et al. 2006); considering the prominent population of Kiss1 neurons at this site and its striking sexual dimorphism, it is worthy to explore whether alterations in the maturation of Kiss1 system are involved in this phenomenon. Similarly, specific mechanistic studies targeting the Kiss1 system need to be implemented as to provide the basis for some of the suspected clinical phenotypes of reproductive endocrine disruption. For instance, recent preliminary evidence suggests that early (gestational) exposures to BPA can up-regulate kisspeptin content at the hypothalamus in adult mice (Panzica et al. 2009); specific studies monitoring earlier changes (i.e., during the pubertal transition) might help to define whether and how a similar phenomenon might contribute to the trend of advancement of the age of puberty in girls, especially if internationally adopted, as detected in different studies in Europe and USA (Bourguignon et al. 2010).

Another interesting question, from a mechanistic perspective, arising from the data summarized in previous sections is the molecular basis for the observed organizing changes detected in Kiss1 neurons following inappropriate exposures to sex steroid-acting compounds. In principle, different mechanisms, including differential epigenetic regulation as well as regulation of cell viability/apoptosis, might contribute to the sexual dimorphism of Kiss1 neuronal populations detected in physiological conditions, and its alteration following estrogenic or androgenic exposures during early development. While this issue had been only superficially addressed, a recent report has evaluated the contribution of BAX-mediated cell death in the development of Kiss1 neurons at the ARC and AVPV, and its potential implication in the sexual dimorphism of these populations, which is especially prominent at the AVPV. Studies in BAX KO mice revealed that females had a larger Kiss1 population at the AVPV, irrespective of the presence of BAX. However, BAX-mediated apoptosis appears critical to shape Kiss1 neuronal population at the ARC (Semaan et al. 2010). In this context, it would be interesting to evaluate whether such BAX-dependent mechanism might be sensitive to the effects of sex steroids, especially at early windows of development.

Other potential source of mechanistic variability may come from the recently-recognized phenomenon that both estrogen-responsive-element (ERE)-dependent and – independent mechanisms are involved in the activational regulation of *Kiss1* gene expression by estrogen. Thus, while all Kiss1 neurons have been shown to abundantly express ERα, recent studies have demonstrated that the stimulatory actions of estradiol on *Kiss1* mRNA levels at the AVPV are conducted through a classical ERE pathway whereas its inhibitory effects on *Kiss1* expression at the ARC do not appear to require such a classical pathway, suggesting alternative regulatory mechanisms (either ERE-independent genomic actions or non-nuclear receptor mediated events) (Gottsch et al. 2009). Admittedly, the potential contribution of such differential signaling to the developmental actions of endogenous (and eventual exogenous) sex steroids has not been evaluated yet. However, such divergent regulatory pathways may help to explain the different sensitivity of Kiss1 neurons to the organizing effects of sex steroids in different hypothalamic nuclei (Kauffman et al. 2007).

Finally, besides xeno-hormones, different environmental cues have been proven to impact the expression and/or function of hypothalamic Kiss1. Among those, the nutritional state of the organism is known to influence pubertal timing and fertility at least partially through the modulation of Kiss1 (Castellano et al. 2009). Intriguingly, EDC have been recently proposed as potential modifiers of body weight as well, and endocrine disruption has been hypothesized as potential contributor to the increased prevalence of obesity and metabolic syndrome in human populations (Diamanti-Kandarakis et al. 2009). Whether early exposures to EDC might have a combined impact on body weight and reproductive maturation and function remains to be solved. Of note, our preliminary studies suggest that postnatal overfeeding in rats causes a persistent increase in body weight and acceleration of puberty that is linked to enhanced expression of Kiss1/kisspeptin at the hypothalamus, whereas the opposite trends (delayed puberty; decreased Kiss1/kisspeptin levels at the hypothalamus) are detected in female rats subjected to postnatal underfeeding (Castellano et al. 2010). In this context, it will be interesting to determine whether neonatally over- or under-weighed animals are more prone to disruption (either in terms of precocious activation or inhibition) of normal development of hypothalamic Kiss1 system by exposures to compounds with sex steroid-like activity. Similarly, the influence of early exposures to EDC in the generation of reproductive diseases, such as polycystic ovarian syndrome (PCOS), linked to severe metabolic complications warrants specific investigation (Diamanti-Kandarakis et al. 2009).

Acknowledgments The author is indebted with Leonor Pinilla, Victor M. Navarro, Miguel A. Sanchez-Garrido and the other members of the research team at the Physiology Section of the University of Cordoba, who actively participated in the generation of experimental data discussed herein. The work from the author's laboratory reviewed in this article was supported by grant BFU 2008-00984 (Ministerio de Ciencia e Innovación, Spain); Project PI042082 (Ministerio de Sanidad, Spain); Project P08-CVI-03788 (Junta de Andalucía, Spain) and EU research contracts EDEN QLK4-CT-2002-00603 and DEER FP7-ENV-2007-1. CIBER Fisiopatología de la Obesidad y Nutrición is an initiative of Instituto de Salud Carlos III, Ministerio de Sanidad, Spain.

References

Adewale HB, Jefferson WN, Newbold RR, Patisaul HB (2009) Neonatal bisphenol-a exposure alters rat reproductive development and ovarian morphology without impairing activation of gonadotropin-releasing hormone neurons. Biol Reprod 81(4):690–699

Adewale HB, Todd KL, Mickens JA, Patisaul HB (2011) The impact of neonatal bisphenol-A exposure on sexually dimorphic hypothalamic nuclei in the female rat. Neurotoxicology 32(1):38–49

Bakker J, Pierman S, Gonzalez-Martinez D (2010) Effects of aromatase mutation (ArKO) on the sexual differentiation of kisspeptin neuronal numbers and their activation by same versus opposite sex urinary pheromones. Horm Behav 57(4–5):390–395

Bateman HL, Patisaul HB (2008) Disrupted female reproductive physiology following neonatal exposure to phytoestrogens or estrogen specific ligands is associated with decreased GnRH activation and kisspeptin fiber density in the hypothalamus. Neurotoxicology 29(6):988–997

Bellingham M, Fowler PA, Amezaga MR, Rhind SM, Cotinot C, Mandon-Pepin B, Sharpe RM, Evans NP (2009) Exposure to a complex cocktail of environmental endocrine-disrupting compounds disturbs the kisspeptin/GPR54 system in ovine hypothalamus and pituitary gland. Environ Health Perspect 117(10):1556–1562

Bellingham M, Fowler PA, Amezaga MR, Whitelaw CM, Rhind SM, Cotinot C, Mandon-Pepin B, Sharpe RM, Evans NP (2010) Foetal hypothalamic and pituitary expression of gonadotrophin-releasing hormone and galanin systems is disturbed by exposure to sewage sludge chemicals via maternal ingestion. J Neuroendocrinol 22(6):527–533

Billings HJ, Connors JM, Altman SN, Hileman SM, Holaskova I, Lehman MN, McManus CJ, Nestor CC, Jacobs BH, Goodman RL (2010) Neurokinin B acts via the neurokinin-3 receptor in the retrochiasmatic area to stimulate luteinizing hormone secretion in sheep. Endocrinology 151(8):3836–3846

Bourguignon JP, Rasier G, Lebrethon MC, Gerard A, Naveau E, Parent AS (2010) Neuroendocrine disruption of pubertal timing and interactions between homeostasis of reproduction and energy balance. Mol Cell Endocrinol 324(1–2):110–120

Castellano JM, Navarro VM, Fernandez-Fernandez R, Castano JP, Malagon MM, Aguilar E, Dieguez C, Magni P, Pinilla L, Tena-Sempere M (2006) Ontogeny and mechanisms of action for the stimulatory effect of kisspeptin on gonadotropin-releasing hormone system of the rat. Mol Cell Endocrinol 257–258:75–83

Castellano JM, Roa J, Luque RM, Dieguez C, Aguilar E, Pinilla L, Tena-Sempere M (2009) KiSS-1/kisspeptins and the metabolic control of reproduction: Physiologic roles and putative physiopathological implications. Peptides 30(1):139–145

Castellano JM, Bentsen AH, Sanchez-Garrido MA, Ruiz-Pino F, Romero M, Pineda R, Garcia-Galiano D, Aguilar E, Pinilla L, Dieguez C, Mikkelsen J, Tena-Sempere M (2010) Early metabolic programming of puberty onset: impact of changes in perinatal feeding on the timing of puberty and the development of hypothalamic kisspeptin system. The Endocrine Society San Diego, USA, pp P2–272

Cheng G, Coolen LM, Padmanabhan V, Goodman RL, Lehman MN (2010) The kisspeptin/neurokinin B/dynorphin (KNDy) cell population of the arcuate nucleus: sex differences and effects of prenatal testosterone in sheep. Endocrinology 151(1):301–311

Christian CA, Moenter SM (2010) The neurobiology of preovulatory and estradiol-induced gonadotropin-releasing hormone surges. Endocr Rev 31(4):544–577

Ciofi P, Lapirot OC, Tramu G (2007) An androgen-dependent sexual dimorphism visible at puberty in the rat hypothalamus. Neuroscience 146(2):630–642

Clarkson J, Herbison AE (2006) Postnatal development of kisspeptin neurons in mouse hypothalamus; sexual dimorphism and projections to gonadotropin-releasing hormone neurons. Endocrinology 147(12):5817–5825

Clarkson J, Boon WC, Simpson ER, Herbison AE (2009) Postnatal development of an estradiol-kisspeptin positive feedback mechanism implicated in puberty onset. Endocrinology 150(7):3214–3220

de d'Anglemont Tassigny X, Colledge WH (2010) The role of kisspeptin signaling in reproduction. Physiology (Bethesda) 25(4):207–217

de d'Anglemont Tassigny X, Fagg LA, Dixon JP, Day K, Leitch HG, Hendrick AG, Zahn D, Franceschini I, Caraty A, Carlton MB, Aparicio SA, Colledge WH (2007) Hypogonadotropic hypogonadism in mice lacking a functional Kiss1 gene. Proc Natl Acad Sci USA 104(25):10714–10719

de Roux N, Genin E, Carel JC, Matsuda F, Chaussain JL, Milgrom E (2003) Hypogonadotropic hypogonadism due to loss of function of the Kiss1-derived peptide receptor GPR54. Proc Natl Acad Sci USA 100(19):10972–10976

Diamanti-Kandarakis E, Bourguignon JP, Giudice LC, Hauser R, Prins GS, Soto AM, Zoeller RT, Gore AC (2009) Endocrine-disrupting chemicals: an endocrine society scientific statement. Endocr Rev 30(4):293–342

Dickerson SM, Guevara E, Woller MJ, Gore AC (2009) Cell death mechanisms in GT1-7 GnRH cells exposed to polychlorinated biphenyls PCB74, PCB118, and PCB153. Toxicol Appl Pharmacol 237(2):237–245

Ebling FJ (2005) The neuroendocrine timing of puberty. Reproduction 129(6):675–683

Evuarherhe O, Leggett JD, Waite EJ, Kershaw YM, Atkinson HC, Lightman SL (2009) Organizational role for pubertal androgens on adult hypothalamic-pituitary-adrenal sensitivity to testosterone in the male rat. J Physiol 587(Pt 12):2977–2985

Fink G (2000) Neuroendocrine regulation of pituitary function: general principles. In: Conn PM, Freeman ME (eds) Neuroendocrinology in physiology and medicine. Humana Press, New Jersey, pp 107–134

Gonzalez-Martinez D, De Mees C, Douhard Q, Szpirer C, Bakker J (2008) Absence of gonadotropin-releasing hormone 1 and Kiss1 activation in alpha-fetoprotein knockout mice: prenatal estrogens defeminize the potential to show preovulatory luteinizing hormone surges. Endocrinology 149(5):2333–2340

Gore AC (2008a) Developmental programming and endocrine disruptor effects on reproductive neuroendocrine systems. Front Neuroendocrinol 29(3):358–374

Gore AC (2008b) Neuroendocrine systems as targets for environmental endocrine-disrupting chemicals. Fertil Steril 89(2 Suppl):e101–e102

Gottsch ML, Navarro VM, Zhao Z, Glidewell-Kenney C, Weiss J, Jameson JL, Clifton DK, Levine JE, Steiner RA (2009) Regulation of Kiss1 and dynorphin gene expression in the murine brain by classical and nonclassical estrogen receptor pathways. J Neurosci 29(29):9390–9395

Guillette LJ Jr (2006) Endocrine disrupting contaminants – beyond the dogma. Environ Health Perspect 114(Suppl 1):9–12

Han SK, Gottsch ML, Lee KJ, Popa SM, Smith JT, Jakawich SK, Clifton DK, Steiner RA, Herbison AE (2005) Activation of gonadotropin-releasing hormone neurons by kisspeptin as a neuroendocrine switch for the onset of puberty. J Neurosci 25(49):11349–11356

Herbison AE (2008) Estrogen positive feedback to gonadotropin-releasing hormone (GnRH) neurons in the rodent: The case for the rostral periventricular area of the third ventricle (RP3V). Brain Res Rev 57:277–287

Herbison AE, de Tassigny X, Doran J, Colledge WH (2010) Distribution and postnatal development of Gpr54 gene expression in mouse brain and gonadotropin-releasing hormone neurons. Endocrinology 151(1):312–321

Homma T, Sakakibara M, Yamada S, Kinoshita M, Iwata K, Tomikawa J, Kanazawa T, Matsui H, Takatsu Y, Ohtaki T, Matsumoto H, Uenoyama Y, Maeda K, Tsukamura H (2009) Significance of neonatal testicular sex steroids to defeminize anteroventral periventricular kisspeptin neurons and the GnRH/LH surge system in male rats. Biol Reprod 81(6):1216–1225

Hrabovszky E, Ciofi P, Vida B, Horvath MC, Keller E, Caraty A, Bloom SR, Ghatei MA, Dhillo WS, Liposits Z, Kallo I (2010) The kisspeptin system of the human hypothalamus: sexual dimorphism and relationship with gonadotropin-releasing hormone and neurokinin B neurons. Eur J Neurosci 31(11):1984–1998

Kauffman AS, Gottsch ML, Roa J, Byquist AC, Crown A, Clifton DK, Hoffman GE, Steiner RA, Tena-Sempere M (2007) Sexual differentiation of Kiss1 gene expression in the brain of the rat. Endocrinology 148(4):1774–1783

Kelce WR, Gray LE (1999) Environmental antiandrogens as endocrine disruptors. In: Naz RK (ed) Endocrine disruptors. Effects on male and female reproductive systems. CRC Press, Boca Raton, pp 247–277

Kotani M, Detheux M, Vandenbogaerde A, Communi D, Vanderwinden JM, Le Poul E, Brezillon S, Tyldesley R, Suarez-Huerta N, Vandeput F, Blanpain C, Schiffmann SN, Vassart G, Parmentier M (2001) The metastasis suppressor gene KiSS-1 encodes kisspeptins, the natural ligands of the orphan G protein-coupled receptor GPR54. J Biol Chem 276(37):34631–34636

Lehman MN, Coolen LM, Goodman RL (2010) Minireview: kisspeptin/neurokinin B/dynorphin (KNDy) cells of the arcuate nucleus: a central node in the control of gonadotropin-releasing hormone secretion. Endocrinology 151(8):3479–3489

Losa SM, Todd KL, Sullivan AW, Cao J, Mickens JA, Patisaul HB (2011) Neonatal exposure to genistein adversely impacts the ontogeny of hypothalamic kisspeptin signaling pathways and ovarian development in the peripubertal female rat. Reprod Toxicol 31(3):280–289

MacLusky NJ, Naftolin F (1981) Sexual differentiation of the central nervous system. Science 211(4488):1294–1302

Morris JA, Jordan CL, Breedlove SM (2004) Sexual differentiation of the vertebrate nervous system. Nat Neurosci 7(10):1034–1039

Muir AI, Chamberlain L, Elshourbagy NA, Michalovich D, Moore DJ, Calamari A, Szekeres PG, Sarau HM, Chambers JK, Murdock P, Steplewski K, Shabon U, Miller JE, Middleton SE, Darker JG, Larminie CG, Wilson S, Bergsma DJ, Emson P, Faull R, Philpott KL, Harrison DC (2001) AXOR12, a novel human G protein-coupled receptor, activated by the peptide KiSS-1. J Biol Chem 276(31):28969–28975

Navarro VM, Tena-Sempere M (2008) The KiSS-1/GPR54 system: putative target for endocrine disruption of reproduction at hypothalamic-pituitary unit? Int J Androl 31(2):224–232

Navarro VM, Castellano JM, Fernandez-Fernandez R, Barreiro ML, Roa J, Sanchez-Criado JE, Aguilar E, Dieguez C, Pinilla L, Tena-Sempere M (2004) Developmental and hormonally regulated messenger ribonucleic acid expression of KiSS-1 and its putative receptor, GPR54, in rat hypothalamus and potent luteinizing hormone-releasing activity of KiSS-1 peptide. Endocrinology 145(10):4565–4574

Navarro VM, Sanchez-Garrido MA, Castellano JM, Roa J, Garcia-Galiano D, Pineda R, Aguilar E, Pinilla L, Tena-Sempere M (2009) Persistent impairment of hypothalamic KiSS-1 system after exposures to estrogenic compounds at critical periods of brain sex differentiation. Endocrinology 150(5):2359–2367

Navarro VM, Castellano JM, McConkey SM, Pineda R, Ruiz-Pino F, Pinilla L, Clifton DK, Tena-Sempere M, Steiner RA (2011) Interactions between Kisspeptin and Neurokinin B in the control of GnRH secretion in the female rat. Am J Physiol Endocrinol Metab 300(1):202–210

Norgil Damgaard I, Main KM, Toppari J, Skakkebaek NE (2002) Impact of exposure to endocrine disrupters in utero and in childhood on adult reproduction. Best Pract Res Clin Endocrinol Metab 16(2):289–309

Oakley AE, Clifton DK, Steiner RA (2009) Kisspeptin signaling in the brain. Endocr Rev 30(6):713–743

Ohtaki T, Shintani Y, Honda S, Matsumoto H, Hori A, Kanehashi K, Terao Y, Kumano S, Takatsu Y, Masuda Y, Ishibashi Y, Watanabe T, Asada M, Yamada T, Suenaga M, Kitada C, Usuki S, Kurokawa T, Onda H, Nishimura O, Fujino M (2001) Metastasis suppressor gene KiSS-1 encodes peptide ligand of a G-protein-coupled receptor. Nature 411(6837):613–617

Ojeda SR, Lomniczi A, Mastronardi C, Heger S, Roth C, Parent AS, Matagne V, Mungenast AE (2006) Minireview: the neuroendocrine regulation of puberty: is the time ripe for a systems biology approach? Endocrinology 147(3):1166–1174

Ojeda SR, Dubay C, Lomniczi A, Kaidar G, Matagne V, Sandau US, Dissen GA (2010) Gene networks and the neuroendocrine regulation of puberty. Mol Cell Endocrinol 324(1–2):3–11

Panzica GC, Mura E, Miceli D, Martini MA, Gotti S, Viglietti-Panzica C (2009) Effects of xenoestrogens on the differentiation of behaviorally relevant neural circuits in higher vertebrates. Ann N Y Acad Sci 1163:271–278

Patisaul HB, Fortino AE, Polston EK (2006) Neonatal genistein or bisphenol-A exposure alters sexual differentiation of the AVPV. Neurotoxicol Teratol 28(1):111–118

Patisaul HB, Todd KL, Mickens JA, Adewale HB (2009) Impact of neonatal exposure to the ERalpha agonist PPT, bisphenol-A or phytoestrogens on hypothalamic kisspeptin fiber density in male and female rats. Neurotoxicology 30(3):350–357

Pineda R, Garcia-Galiano D, Roseweir A, Romero M, Sanchez-Garrido MA, Ruiz-Pino F, Morgan K, Pinilla L, Millar RP, Tena-Sempere M (2010) Critical roles of kisspeptins in female puberty and preovulatory gonadotropin surges as revealed by a novel antagonist. Endocrinology 151(2):722–730

Pinilla L, Tena-Sempere M, Gonzalez D, Aguilar E (1995) Mechanisms of altered LH secretion in neonatally oestrogenized male rats. J Endocrinol 147(1):43–50

Popa SM, Clifton DK, Steiner RA (2008) The role of Kisspeptins and GPR54 in the neuroendocrine regulation of reproduction. Annu Rev Physiol 70:213–238

Ramaswamy S, Seminara SB, Ali B, Ciofi P, Amin NA, Plant TM (2010) Neurokinin B stimulates GnRH release in the male monkey (*Macaca mulatta*) and is colocalized with Kisspeptin in the arcuate nucleus. Endocrinology 151(9):4494–4503

Rance NE (2009) Menopause and the human hypothalamus: evidence for the role of kisspeptin/neurokinin B neurons in the regulation of estrogen negative feedback. Peptides 30(1):111–122

Rasier G, Parent AS, Gerard A, Lebrethon MC, Bourguignon JP (2007) Early maturation of gonadotropin-releasing hormone secretion and sexual precocity after exposure of infant female rats to estradiol or dichlorodiphenyltrichloroethane. Biol Reprod 77(4):734–742

Richard N, Galmiche G, Corvaisier S, Caraty A, Kottler ML (2008) KiSS-1 and GPR54 genes are co-expressed in rat gonadotrophs and differentially regulated in vivo by oestradiol and gonadotrophin-releasing hormone. J Neuroendocrinol 20(3):381–393

Roa J, Aguilar E, Dieguez C, Pinilla L, Tena-Sempere M (2008) New frontiers in kisspeptin/GPR54 physiology as fundamental gatekeepers of reproductive function. Front Neuroendocrinol 29(1):48–69

Semaan SJ, Murray EK, Poling MC, Dhamija S, Forger NG, Kauffman AS (2010) BAX-dependent and BAX-independent regulation of Kiss1 neuron development in mice. Endocrinology 151(12):5807–5817

Seminara SB, Messager S, Chatzidaki EE, Thresher RR, Acierno JS Jr, Shagoury JK, Bo-Abbas Y, Kuohung W, Schwinof KM, Hendrick AG, Zahn D, Dixon J, Kaiser UB, Slaugenhaupt SA, Gusella JF, O'Rahilly S, Carlton MB, Crowley WF Jr, Aparicio SA, Colledge WH (2003) The GPR54 gene as a regulator of puberty. N Engl J Med 349(17):1614–1627

Shahab M, Mastronardi C, Seminara SB, Crowley WF, Ojeda SR, Plant TM (2005) Increased hypothalamic GPR54 signaling: a potential mechanism for initiation of puberty in primates. Proc Natl Acad Sci USA 102(6):2129–2134

Sikka SC, Naz RK (1999) Endocrine disruptors and male infertility. In: Naz RK (ed) Endocrine disruptors. Effects on male and female reproductive systems. CRC Press, Boca Raton, pp 225–246

Skakkebaek NE, Jorgensen N, Main KM, Rajpert-De Meyts E, Leffers H, Andersson AM, Juul A, Carlsen E, Mortensen GK, Jensen TK, Toppari J (2006) Is human fecundity declining? Int J Androl 29(1):2–11

Tena-Sempere M (2010a) Kisspeptin/GPR54 system as potential target for endocrine disruption of reproductive development and function. Int J Androl 33(2):360–368

Tena-Sempere M (2010b) Roles of Kisspeptins in the control of hypothalamic-gonadotropic function: Focus on sexual differentiation and puberty onset. Endocr Dev 17:52–62

Tena-Sempere M, Huhtaniemi I (2003) Gonadotropins and gonadotropin receptors. In: Fauser BCJM (ed) Reproductive medicine – Molecular, cellular and genetic fundamentals. Parthenon Publishing, New York, pp 225–244

Tena-Sempere M, Pinilla L, Gonzalez LC, Aguilar E (2000) Reproductive disruption by exposure to exogenous estrogenic compounds during sex differentiation: lessons from the neonatally estrogenized male rat. Curr Top Steroid Res 3:23–37

Topaloglu AK, Reimann F, Guclu M, Yalin AS, Kotan LD, Porter KM, Serin A, Mungan NO, Cook JR, Ozbek MN, Imamoglu S, Akalin NS, Yuksel B, O'Rahilly S, Semple RK (2009) TAC3 and TACR3 mutations in familial hypogonadotropic hypogonadism reveal a key role for Neurokinin B in the central control of reproduction. Nat Genet 41(3):354–358

Toppari J, Larsen JC, Christiansen P, Giwercman A, Grandjean P, Guillette LJ Jr, Jegou B, Jensen TK, Jouannet P, Keiding N, Leffers H, McLachlan JA, Meyer O, Muller J, Rajpert-De Meyts E, Scheike T, Sharpe R, Sumpter J, Skakkebaek NE (1996) Male reproductive health and environmental xenoestrogens. Environ Health Perspect 104(Suppl 4):741–803

Wadas BC, Hartshorn CA, Aurand ER, Palmer JS, Roselli CE, Noel ML, Gore AC, Veeramachaneni DN, Tobet SA (2010) Prenatal exposure to vinclozolin disrupts selective aspects of the gonadotrophin-releasing hormone neuronal system of the rabbit. J Neuroendocrinol 22(6):518–526

Wakabayashi Y, Nakada T, Murata K, Ohkura S, Mogi K, Navarro VM, Clifton DK, Mori Y, Tsukamura H, Maeda K, Steiner RA, Okamura H (2010) Neurokinin B and dynorphin A in kisspeptin neurons of the arcuate nucleus participate in generation of periodic oscillation of neural activity driving pulsatile gonadotropin-releasing hormone secretion in the goat. J Neurosci 30(8):3124–3132

Effects of Prenatal Exposure to Endocrine Disrupters on Cerebral Cortex Development

Anne-Simone Parent, Elise Naveau, and Jean-Pierre Bourguignon

Abstract For several decades, the focus of most studies on endocrine disrupting chemicals (EDCs) has been the reproductive system, with fertility and hormone-dependent cancers being the most critical issues.

Cerebral cortex development is very sensitive to hormonal environment, in particular thyroid hormones and sex steroids. Experimental data concerning early exposure to polychlorinated biphenyls (PCBs) illustrate the detrimental effect of endocrine disrupters on the central nervous system. While epidemiological studies have reported a negative correlation between prenatal exposure to PCBs and cognitive performances, the molecular and cellular mechanisms of such neurotoxicity are incompletely understood. This paper will review the role of thyroid hormones and sex steroids in cerebral cortex development and will illustrate, with PCBs and bisphenol A, the potential effects of EDCs on cerebral cortex development.

Introduction

For a long time, most studies have focused on the peripheral effects of endocrine disrupting chemicals (EDCs) on the testis and ovary (review in Diamanti-Kandarakis et al. 2009) as well as their effects on sex steroid-sensitive peripheral structures, such as the prostate or breast (Diamanti-Kandarakis et al. 2009). However, sex steroids, corticoids and thyroid hormones play a key role in the development of the central nervous system and of the cerebral cortex in particular. The critical role of these hormonal systems explains the sensitivity of the cerebral cortex to EDCs. These brain regions are complex networks of neurons and surrounding

A.-S. Parent (✉)
Developmental Neuroendocrinology Unit, GIGA-Neurosciences, University of Liège, Liège, Belgium
e-mail: asparent@ulg.ac.be

J.-P. Bourguignon et al. (eds.), *Multi-System Endocrine Disruption*,
Research and Perspectives in Endocrine Interactions,
DOI 10.1007/978-3-642-22775-2_3, © Springer-Verlag Berlin Heidelberg 2011

glial cells, which are modulated by paracrine or autocrine neurotransmitters as well as peripheral hormones and chemicals produced in the body or in the environment. Hormones have lifelong effects on central functions by influencing cellular proliferation, dendritic outgrowth, synaptogenesis or neurotransmitter secretion. Structural changes in the brain following hormonal alterations during fetal and perinatal life result in functional consequences in adolescence and adulthood. Typical examples are anovulation and infertility after perinatal exposure to sex steroids (Sawaki et al. 2003) and cognitive dysfunction after fetal hypothyroidism (DeLange 2000). As is the case for other systems, the developing cerebral cortex seems particularly sensitive to endocrine disruption. It is known that some neurological diseases are explained by an alteration of early processes such as progenitor proliferation, migration or differentiation. We could face a similar pattern in the case of EDCs, which could affect cortical development and lead to altered cognitive function later in life. This review will focus on the effects of EDCs on cerebral cortex development.

Developmental Processes in the Cortex Regulated by Thyroid Hormones and Sex Steroids

Knowing the developmental processes that depend on sex steroids and thyroid hormones, one can hypothesize on the stages potentially affected by EDCs. We will review here the multiple actions of thyroid hormones and sex steroids on cortical development. Estradiol is a possible factor promoting the development, function and survival of neurons (McEwen and Alves 1999) through classical genomic interactions with the nuclear estrogen receptor (ER) and also non-genomic interactions with membrane receptors. Neurons, astrocytes and neuronal progenitors express ERs. In particular, astrocytes influence neural development in part by synthesizing estrogens (Garcia-Segura and Melcangi 2006). Interestingly, alpha-fetoprotein (AFP) is expressed at high levels in radial glial cells but at lower levels by intermediate progenitors. Thus high levels of AFP in the ventricular zone could inhibit E2 (17-beta Estradiol)-promoted proliferation in this region whereas low levels of AFP in the subventricular zone could allow a stronger effect of E2 on intermediate progenitors (Martinez-Cerdeno et al. 2006). Estrogens also stimulate neurogenesis in adult rodents and increase proliferation in cortical progenitor cells by shortening the G1 phase (Martinez-Cerdeno et al. 2006). Because EDCs can affect the ER directly or indirectly through estrogen biosynthesis or metabolism, it is important that studies of the action of EDCs examine those different structures and functions in the cortex

During fetal and neonatal life, neuronal and glial proliferation, migration, and differentiation depend on thyroid hormones. Thyroid hormone action is mediated by two classes of nuclear receptors (Forrest and Vennstrom 2000) that exhibit differential spatial and temporal expression in the brain, suggesting that thyroid hormones have multiple functions during brain development (Horn and Heuer 2010).

Thyroid hormone receptors are expressed in neurons, astrocytes, and oligodendrocytes and precursors before the fetal thyroid is functional, suggesting a role for hormones of maternal origin. Triiodothyronine (T3) regulates the expression of genes coding for the growth factors, cell surface receptors and transcription factors involved in cell cycle regulation and proliferation (reviewed in Puzianowska-Kuznicka et al. 2006). The action of T3 is not homogeneous and depends on the cell type and its developmental state. T3 blocks proliferation and induces differentiation of oligodendrocyte progenitor cells (Baas et al. 1997). This effect results from a rapid decrease of the transcription factor E2F1 in oligodendrocyte precursors, which induces a decrease of proliferation by arresting the cells in G1 and S phases (Nygard et al. 2003). Tokumoto et al. (2001) also showed that thyroid hormones promote oligodendrocyte differentiation through another pathway involving p53 proteins. In addition to these few studies suggesting a role for thyroid hormones on cell proliferation in the cortex, several studies have reported an effect on cell migration and differentiation. For example, T4 promotes actin polymerization through non-genomic action in developing neurons (reviewed in Cheng et al. 2010). Actin polymerization is necessary to recognize the laminin guidance molecule during migration (Farwell et al. 2005). Thyroid hormones also regulate the organization of the actin cytoskeleton in astrocytes during development, thus affecting the production and deposition of laminin at the surface of astrocytes that is necessary for neuronal migration (Farwell and Dubord-Tomasetti 1999). In ex vivo studies, maternal hypothyroxinemia alters radial and tangential neuronal migration (Lavado-Autric et al. 2003; Auso et al. 2004). In these experiments, green fluorescent protein-medial ganglionic eminence (GFP-MGE)- derived neurons from hypothyroxinemic mothers showed a normal migratory behavior whereas GFP-MGE-neurons from normal or hypothyroxinemic mothers showed disrupted migration when explanted into the neocortex of embryos from hypothyroxinemic dams. These studies suggest an effect caused not by the migratory neurons themselves but by elements guiding the migration (Cuevas et al. 2005). Thyroid hormones also regulate the expression and distribution of molecules, such as actin or tenascin (Farwell et al. 2005; Alvarez-Dolado et al. 1998), that interact with the extracellular matrix and facilitate neurite outgrowth. Overall, these examples illustrate that thyroid hormones are involved in multiple aspects of early brain development including proliferation, differentiation and migration of progenitors. Disruption of thyroid function by EDCs such as PCBs could thus cause neurological deficits that are very similar to hypothyroidism.

Epidemiological Data

As stated above, thyroid hormones and sex steroids play a fundamental role in the development of the cerebral cortex, and many environmental chemicals are able to alter thyroid function or sex steroid action. One example is a group of chemicals called polychlorinated biphenyls (PCBs), which is a family of 209 different congeners used in lubricating oils and plasticizers. Because of their long half-life (Ogura et al. 2005), they are still ubiquitous environmental contaminants, found in high concentrations in humans and animals, even though they have been banned in Europe and the USA since the 1970s. The first observation pointing to the neurotoxic effects of PCBs followed an accidental exposure in Taiwan in which children exposed in utero showed impaired cognitive function at 5 years of age (McKinney and Waller 1994). The major difficulty in such studies is the long delay between the exposure and its measurable effect. Several other follow-up studies have shown a negative correlation between in utero exposure to PCBs and cognitive performance and memory in infants and children (reviewed in Schantz et al. 2003). Those results are consistent with observations made in rodents. It is interesting to note that the levels of exposure in recent studies are lower than in earlier studies but still negatively correlate with cognitive function. More recent studies are developing analytic methods to correlate neurodevelopmental toxicity with specific congeners. Some PCB congeners could lower thyroid hormones levels in serum and thus induce a state similar to hypothyroidism. However, epidemiological studies reported levels of T3, T4 and thyroid-stimulating hormone (TSH) that were in the normal range in pregnant women and newborns. But some studies reported that higher levels of PCBs in maternal and cord blood were associated with higher levels of TSH in newborns.

Bisphenol A (BPA) is a ubiquitous industrial chemical used in the manufacture of plastics and epoxy resins. It is present in many plastic bottles, baby bottles and food cans and is found in the urine of more than 90% of Americans (Melzer et al. 2010). Fetuses and newborns seem to be particularly exposed since BPA is known to cross the placenta and is found in high concentration in amniotic fluids and cord serum (Schönfelder et al. 2002). BPA is a weak estrogenic compound binding to ERα and β as well as membrane receptors. It also antagonizes T3 activation of the thyroid receptor, and developmental exposure to BPA induces a state similar to thyroid hormone resistance. Very little data are available concerning the effects of BPA on cerebral function in human. One caveat for those studies is that virtually everybody in the Western world has been exposed to BPA. However, knowing that BPA alters sex steroid and thyroid hormone function, one can hypothesize that perinatal exposure to BPA could lead to alterations in cerebral cortex development. Nakagami et al. (2009) reported an alteration in male behavior toward mothers after prenatal exposure to BPA in non-human primates, which suggests that, because of its interaction with sex steroid receptors, BPA could alter the sexual differentiation of the brain taking place perinatally.

Molecular Mechanisms of Endocrine Disruption of Cerebral Cortex Development

The molecular mechanisms by which EDCs can cause alterations of cerebral cortex development are still incompletely understood, but some data are available for PCBs. Some studies suggest that PCBs cause a state of relative hypothyroidism that could explain their neurotoxicity. Interestingly, the cerebellum and the auditory system that depend on thyroid hormone are very sensitive to PCBs (Koibuchi and Chin 2000). However, PCBs do not only produce effects consistent with hypothyroidism, since the expressions of some thyroid hormone-responsive genes are increased after neonatal exposure to PCBs (Gauger et al. 2004). Some in vitro studies have shown that some PCBs congeners act as thyroid hormone receptors agonists (Fritsche et al. 2005). Based on their chemical structure, PCBs can act through different pathways (McKinney and Waller 1994). Coplanar congeners bind to cytosolic aryl hydrocarbon receptors (AhR), a ligand-dependent transcription factor involved in cell proliferation and differentiation (Dietrich and Kaina 2010). However, the neurotoxic effects of PCBs on development might not be entirely explained by AhR. Some PCBs can also alter neurotransmission and intracellular signaling (Kodavanti 2006).

BPA is another example of the complexity of the mechanisms of action of EDCs on the brain. Some studies suggest that BPA could indeed affect cerebral cortex development. Prenatal exposure to BPA does not affect progenitor cell proliferation in mice but it alters the number of cells in each of the cortical layers postnatally (Nakamura et al. 2007). It has also been shown that BPA could have an antiestrogenic action on synaptogenesis in the rodent and non-human primate hippocampus (Hajszan and Leranth 2010; Leranth et al. 2008). BPA is classically known to act as an estrogen agonist with an affinity for ERs that is much lower than estradiol. It is also able to interact with membrane receptors at very low doses. Besides its action on the ER, BPA can also act as a competitive inhibitor for androgen receptors and can disrupt their nuclear localization as well as their trans-activation (review in Wolstenholme et al. 2010). Those actions at the level of the sex steroids receptors could alter the estrogenic/androgenic balance existing in the fetal brain and explain a possible disruption of sexual differentiation of the cerebral cortex. Besides its action on sex steroid receptor, BPA appears to act as a thyroid hormone receptor antagonist in vitro. It blocks T3-dependent oligodendrocyte development (Seiwa et al. 2004) and induces a state similar to thyroid hormone resistance in vivo with increased T4 but did not change THS in exposed animals (Zoeller et al. 2005). Some new mechanisms of action for BPA have been described in the brain. It has recently been shown that low doses of BPA prenatally increase AhR (Nishizawa et al. 2005a) and AhR (Nishizawa et al. 2005b) repressor expression in the brain, suggesting that BPA could affect the cell proliferation and differentiation regulated by AhR. Very recently, perinatal exposure to BPA has been shown to alter methylation of genes involved in prostate cancer (Ho et al. 2006) as well as genes coding for fur color (Dolinoy et al. 2007). But only one study focused on the methylation status of the brain after exposure to BPA. Prenatal exposure to low doses of BPA induced a decreased methylation of two loci, VPS52 and LOC72325, in the brain that correlated with an increased expression of those genes (Yaoi et al. 2008). The function of those genes is not completely understood but the authors hypothesize that changes in their methylation could promote neuronal differentiation and migration.

References

Alvarez-Dolado M, Gonzalez-Sancho JM, Bernal J, Munoz A (1998) Developmental expression of the tenascin-C is altered by hypothyroidism in the rat brain. Neuroscience 84:309–322

Auso E, Lavado-Autric R, Cuevas E, Del Rey FE, Morreale De Escobar G, Berbel P (2004) A moderate and transient deficiency of maternal thyroid function at the beginning of fetal neocorticogenesis alters neuronal migration. Endocrinology 145:4034–4036

Baas D, Bourbeau D, Sarlieve LL, Ittel ME, Dussault JH, Puymirat J (1997) Oligodendrocyte maturation and progenitor cell proliferation are independently regulated by thyroid hormone. Glia 19:324–332

Cheng SY, Leonard JL, Davis PJ (2010) Molecular aspects of thyroid hormone actions. Endocr Rev 31:139–170

Cuevas E, Auso E, Telefont M, Morreale de Escobar G, Sotelo C, Berbel P (2005) Transient maternal hypothyroxinemia at onset of corticogenesis alters tangential migration of medial ganglionic eminence-derived neurons. Eur J Neurosci 22:541–551

DeLange F (2000) Endemic cretinism. In: Braverman LE, Utiger RD (eds) The thyroid. Lippincott, Philadelphia, pp 756–767

Diamanti-Kandarakis E, Bourguignon JP, Giudice LC, Hauser R, Prins GS, Soto AM, Zoeller RT, Gore AC (2009) Endocrine-disrupting chemicals: an Endocrine Society scientific statement. Endocr Rev 30:293–342

Dietrich C, Kaina B (2010) The aryl hydrocarbon receptor (AhR) in the regulation of cell-cell contact and tumor growth. Carcinogenesis 31:1319–1328

Dolinoy DC, Huang D, Jirtle RL (2007) Maternal nutrient supplementation counteracts bisphenol A-induced DNA hypomethylation in early development. Proc Natl Acad Sci USA 104:13056–13061

Farwell AP, Dubord-Tomasetti SA (1999) Thyroid hormone regulates the extracellular organization of laminin on astrocytes. Endocrinology 140:5014–5021

Farwell AP, Dubord-Tomasetti SA, Pietzykowski AZ, Stachelek SJ, Leonard JL (2005) Regulation of cerebellar neuronal migration and neurite outgrowth by thyroxine and 3,3′,5′-triiodothyronine. Brain Res Dev Brain Res 154:121–135

Forrest D, Vennstrom B (2000) Functions of thyroid hormone receptors in mice. Thyroid 10:41–52

Fritsche E, Cline JE, Nguyen NH, Scanlan TS, Abel J (2005) Polychlorinated biphenyls disturb differentiation of normal human neural progenitor cells: clue for involvement of thyroid hormone receptors. Environ Health Perspect 113:871–876

Garcia-Segura LM, Melcangi RC (2006) Steroids and glial cell function. Glia 54:485–498

Gauger KJ, Kato Y, Haraguchi K, Lehmler HJ, Robertson LW, Bansal R, Zoeller RT (2004) Polychlorinated biphenyls (PCBs) exert thyroid hormone-like effects in the fetal rat brain but do not bind to thyroid hormone receptors. Environ Health Perspect 112:516–523

Hajszan T, Leranth C (2010) Bisphenol A interferes with synaptic remodeling. Front Neuroendocrinol 31:519–530

Ho SM, Tang WY, Belmonte de Frausto J, Prins GS (2006) Developmental exposure to estradiol and bisphenol A increases susceptibility to prostate carcinogenesis and epigenetically regulates phosphodiesterase type 4 variant 4. Cancer Res 66:5624–5632

Horn S, Heuer H (2010) Thyroid hormone action during brain development: more questions than answers. Mol Cell Endocrinol 315:19–26

Kodavanti PR (2006) Neurotoxicity of persistent organic pollutants: possible mode(s) of action and further considerations. Dose Response 3:273–305

Koibuchi N, Chin WW (2000) Thyroid hormone action and brain development. Trends Endocrinol Metab 11:123–124

Lavado-Autric R, Auso E, Garcia-Velasco JV, Arufe Mdel C, Escobar del Rey F, Berbel P, Morreale de Escobar G (2003) Early maternal hypothyroxinemia alters histogenesis and cerebral cortex cytoarchitecture of the progeny. J Clin Investig 111:1073–1082

Leranth C, Hajszan T, Szigeti-Buck K, Bober J, MacLusky NJ (2008) Bisphenol A prevents the synaptogenic response to estradiol in hippocampus and prefrontal cortex of ovariectomized nonhuman primates. Proc Natl Acad Sci USA 105:14187–14191

Martinez-Cerdeno V, Noctor SC, Kriegstein AR (2006) Estradiol stimulates progenitor cell division in the ventricular and subventricular zones of the embryonic neocortex. Eur J Neurosci 24:3475–3488

McEwen BS, Alves SE (1999) Estrogen actions in the central nervous system. Endocr Rev 20:279–307

McKinney JD, Waller CL (1994) Polychlorinated biphenyls as hormonally active structural analogues. Environ Health Perspect 102:290–297

Melzer D, Rice NE, Lewis C, Henley WE, Galloway TS (2010) Association of urinary bisphenol a concentration with heart disease: evidence from NHANES 2003/06. PLoS One 5:e8673

Nakagami A, Negishi T, Kawasaki K, Imai N, Nishida Y, Ihara T, Kuroda Y, Yoshikawa Y, Koyama T (2009) Alterations in male infant behaviors towards its mother by prenatal exposure to bisphenol A in cynomolgus monkeys (*Macaca fascicularis*) during early suckling period. Psychoneuroendocrinology 34:1189–1197

Nakamura K, Itoh K, Sugimoto T, Fushiki S (2007) Prenatal exposure to bisphenol A affects adult murine neocortical structure. Neurosci Lett 420:100–105

Nishizawa H, Imanishi S, Manabe N (2005a) Effects of exposure in utero to bisphenol a on the expression of aryl hydrocarbon receptor, related factors, and xenobiotic metabolizing enzymes in murine embryos. J Reprod Dev 51:593–605

Nishizawa H, Morita M, Sugimoto M, Imanishi S, Manabe N (2005b) Effects of in utero exposure to bisphenol A on mRNA expression of arylhydrocarbon and retinoid receptors in murine embryos. J Reprod Dev 51:315–324

Nygard M, Wahlstrom GM, Gustafsson MV, Tokumoto YM, Bondesson M (2003) Hormone-dependent repression of the E2F-1 gene by thyroid hormone receptors. Mol Endocrinol 17:79–92

Ogura I, Gamo M, Masunaga S, Nakanishi J (2005) Quantitative identification of sources of dioxin-like polychlorinated biphenyls in sediments by a factor analysis model and a chemical mass balance model combined with Monte Carlo techniques. Environ Toxicol Chem 24:277–285

Puzianowska-Kuznicka M, Pietrzak M, Turowska O, Nauman A (2006) Thyroid hormones and their receptors in the regulation of cell proliferation. Acta Biochim Polon 53:641–650

Sawaki M, Noda S, Muroi T, Mitoma H, Takakura S, Sakamoto S, Yamasaki K (2003) In utero through lactational exposure to ethinyl estradiol induces cleft phallus and delayed ovarian dysfunction in the offspring. Toxicol Sci 75:402–411

Schantz SL, Widholm JJ, Rice DC (2003) Effects of PCB exposure on neuropsychological function in children. Environ Health Perspect 111:357–376

Schönfelder G, Wittfoht W, Hopp H, Talsness CE, Paul M, Chahoud I (2002) Parent bisphenol A accumulation in the human maternal-fetal-placental unit. Environ Health Perspect 110: A703–A707

Seiwa C, Nakahara J, Komiyama T, Katsu Y, Iguchi T, Asou H (2004) Bisphenol A exerts thyroid hormone-like effects on mouse oligodendrocyte precursor cells. Neuroendocrinology 80:21–30

Tokumoto YM, Tang DG, Raff MC (2001) Two molecularly distinct intracellular pathways to oligodendrocyte differentiation: role of a p53 family protein. EMBO J 20:5261–5268

Wolstenholme JT, Rissman EF, Connelly JJ (2010) The role of bisphenol A in shaping the brain, epigenome and behavior. Horm Behav 59:296–305

Yaoi T, Itoh K, Nakamura K, Ogi H, Fujiwara Y, Fushiki S (2008) Genome-wide analysis of epigenomic alterations in fetal mouse forebrain after exposure to low doses of bisphenol A. Biochem Biophys Res Commun 376:563–567

Zoeller RT, Bansal R, Parris C (2005) Bisphenol-A, an environmental contaminant that acts as a thyroid hormone receptor antagonist in vitro, increases serum thyroxine, and alters RC3/neurogranin expression in the developing rat brain. Endocrinology 146:607–612

Endocrine Disruption of the Thyroid and its Consequences in Development

R. Thomas Zoeller

Abstract Thyroid hormone (TH) is essential for normal human development. This is particularly true for the brain, but it is also true for other organs and systems. Despite the universally held recognition that TH is required for brain development, the specific role of TH in brain development is incompletely understood at best. In part, the difficulty in understanding the role of TH in brain development is due to the complexity of the processes whereby biologically active TH is delivered to target cells and to the complexity of TH action on its receptors. The fetus does not produce sufficient TH for its needs prior to about 20 weeks gestation in the human (about 18 days in the rodent). During this period, a complex interaction of transporters and enzymes is required to deliver TH to the fetal brain. Once at the site of action, the effects of TH are mediated by nuclear receptors that bind to DNA regulatory elements and interact with a complex of other proteins to influence the expression of specific target genes. Studies are beginning to reveal the developmental processes affected by TH and the mechanisms underlying these effects. Given the complexity of TH action on development, it is not surprising that environmental chemicals that interfere with TH action will likewise have complex effects. However, it is essential to recognize that industrial chemicals are being found that influence TH action in unexpected ways. Molecular studies that focused on polychlorinated biphenyls (PCBs), polybrominated biphenyl ethers (PBDEs), bisphenol-A (BPA), and others indicate that a number of chemicals to which the human population is routinely exposed during development can interfere both directly and indirectly with TH action, producing consequences that are not identical to thyroid disease itself. These studies have profound implications for public health and for global strategies to protect the public by applying modern science in regulatory domains charged with chemical safety. In addition, the literature

R.T. Zoeller (✉)
Biology Department and Molecular and Cellular Biology Program, University of Massachusetts Amherst, Amherst, MA, USA
e-mail: tzoeller@bio.umass.edu

J.-P. Bourguignon et al. (eds.), *Multi-System Endocrine Disruption*,
Research and Perspectives in Endocrine Interactions,
DOI 10.1007/978-3-642-22775-2_4, © Springer-Verlag Berlin Heidelberg 2011

describing the environmental effects of perchlorate exposure on the thyroid system may be showing us that short-term, high-dose studies to characterize the risk of adverse health outcome to these exposures are not capable of predicting the risk of long-term, low-dose exposures. Considering the importance of the thyroid system in directing development, it will be essential to clarify these issues.

Introduction

Thyroid hormone (TH) is essential for normal development in humans and animals (Horn and Heuer 2010). The medical and regulatory implications of this observation are tangible in that both clinical guidelines and regulatory decisions are designed to monitor and protect thyroid status of pregnant women and their neonates (EPA 1998; Korada et al. 2008; Rovet and Daneman 2003). However, many questions remain concerning the precise role of TH in brain development and the consequences of TH dysregulation during development. The goal of this short review is to highlight new areas of research that form the basis for a credible case for concern that environmental contaminants can interfere with TH action and pediatric health in ways that are not currently evaluated or recognized, and that reducing contaminant exposures may decrease the pediatric neurobehavioral disease burden.

Pediatric Neurobehavioral Disease Burden

Recent summary statistics of children in the United States indicate that nearly one in six U.S. children, 3–17 years old, are diagnosed with a learning disability or attention deficit (Bloom et al. 2009). This very large disease burden in the pediatric population may have important roots in environmental exposures, although this is a very challenging hypothesis to confirm. Some neurobehavioral disorders, such as Autism spectrum, have been linked to proximity to sources of environmental degradation (Windham et al. 2006) and may be an important indication of an environmental link. The worldwide prevalence of attention deficit disorders is highly variable geographically, but investigators in this area indicate that methodological differences in studies in different sectors of the world make it very difficult to make conclusions about variability related to geographic location and, perhaps, environmental causes (Polanczyk et al. 2007). However, there are substantial differences in neuroanatomical, neurochemical and behavioral components of children diagnosed with ADHD compared to non ADHD controls, and this finding may begin to provide a link to environmental causes (Aguiar et al. 2010). This link (OK?)has been most intensely studied for exposures to polychlorinated biphenyls (PCBs) and lead (Eubig et al. 2010). Overall, these studies indicate a link between PCB or lead body burden and ADHD in both animal and human studies. These are important examples of the ways in which environmental links will be identified for complex and heterogeneous diseases such as ADHD.

There are very few studies to link thyroid dysfunction to environmental contaminants and to generalized neurobehavioral disorders, although we know that a wide variety of chemicals can interfere in various ways with thyroid function or TH action. In this regard, it might be noteworthy that the incidence of congenital hypothyroidism (CH) appears to be increasing (Corbetta et al. 2009; Harris and Pass 2007) because we know that CH is associated with life-long neurobehavioral deficits in this population (Blasi et al. 2009; Dimitropoulos et al. 2009; Kemper et al. 2010; Kugelman et al. 2009; van der Sluijs et al. 2008). Thus, environmental exposures may be a significant contributor to the neurobehavioral disease burden in U.S. children, and although this burden will not be attributed solely to defects in TH signaling, these defects may be important and may not be easily recognized. Molecular biomarkers of TH signaling may be important supplements to biomarkers of thyroid dysregulation (serum hormone levels) to identify these problems. This issue will be developed more fully in this chapter.

Molecular Actions of TH on Brain Development

TH influences brain development in large part by directly regulating target gene expression through the activities of two nuclear transcription factors, TRα and TRβ (McEwan 2009). These TH receptors (TRs) are members of the steroid/thyroid superfamily of hormone-regulated transcription factors and as such exert direct actions on the abundance of specific mRNAs being expressed (Dong et al. 2009). Importantly, TRs bind to DNA in the absence of hormone, and it is this aporeceptor that produces damage to the developing nervous system in hypothyroidism (Wondisford 2003). This result was first observed when the TRs were genetically deleted without causing the same neurodevelopmental phenotype observed when animals were treated with goitrogenic agents (Gothe et al. 1999); in fact, targeted deletion of the TRα gene can protect the animal from the adverse consequences of low TH in the developing cerebellum (Morte et al. 2002). In addition, targeted insertion of a mutation in the TRβ ligand binding domain interferes with TH binding and causes severe impairment of cerebellar development (Hashimoto et al. 2001; Portella et al. 2010), whereas the targeted insertion of a mutation that interferes with TRβ binding to DNA does not produce such damage (Shibusawa et al. 2003). These findings demonstrate that the unliganded TR bound to DNA is the cause of brain damage in the condition of hypothyroidism.

These studies represent only a small proportion of those that investigate the molecular actions of TH in brain development, and they provide important insights into the specific role of the different TR isoforms in brain development (Flamant and Quignodon 2010; Oetting and Yen 2007). These studies identify important molecular details of the actions of TH and its receptors, which will be important as we consider the ways in which environmental chemicals can interfere with TH signaling. In addition, a number of studies have begun to investigate the specific genes regulated by TH in the brain. In principle, the identity of TH-responsive genes in the developing brain may provide insight into the developmental events regulated by TH and the timing of TH actions on those events. For example, our early studies using differential display identified several genes expressed in the fetal rat brain that respond to manipulation of maternal thyroid status before the onset of fetal thyroid function (Dowling et al. 2000, 2001; Dowling and Zoeller 2000). Among these were HES genes (Bansal et al. 2005) that play a role in Notch-directed fate specification (Schuurmans and Guillemot 2002; Soen et al. 2006; Weller et al. 2006), suggesting that an early developmental event targeted by TH action is the differentiation of neural progenitor cells that become either glial elements or neurons. Thyroid hormone also appears to influence fate specification of progenitor cells that give rise to either oligodendrocytes or astrocytes (Billon et al. 2001, 2002), and we have shown that the balance of production of these cell types in the corpus callosum is directly correlated with serum T_4 (Sharlin et al. 2008).

Considering the potentially important role of TH in fate specification, it is critical to recognize that TH effects on the transcriptional profile of the developing brain will reflect both direct actions on genes and the change in cellular phenotype of the brain. That is, when thyroid status is altered in the pregnant animal or in the neonate, the cellular composition of the brain – i.e., the balance of production of different cell types – will become altered, which will be reflected in an altered transcriptional profile. Several transcriptional profiling studies have been published to account for the effects of TH on brain development (Diez et al. 2008; Dong et al. 2005; Kobayashi et al. 2009; Morte et al. 2010a,b; Royland et al. 2008; Takahashi et al. 2008). Although there are significant differences in the profiles of genes characterized by these different studies, there are also many significant differences in the experimental design. For example, the timing, duration and severity of altered thyroid status are likely to be important, as well as the technical approach by which thyroid status is manipulated (e.g., surgery versus chemical thyroidectomy), the hormone employed for replacement (T_4 and/or T_3) and the dose and pattern of hormone supplementation. Moreover, different regions of the brain are evaluated (cortex, striatum, hippocampus, cerebellum) at different times during development. In fact, it is reasonable that these variables would be important, because the role of TH in brain development changes over time and across regions (Zoeller and Rovet 2004).

Targeted studies can also provide important insight. For example, small transient reductions in serum TH can alter neuronal migration in the fetal cerebral cortex (Auso et al. 2004; Lavado-Autric et al. 2003), and detailed studies of reelin expression – a protein important for cortical neuron migration (Rice and Curran 2001) – indicate that TH exerts a direct action on the expression of this gene (Pathak et al. 2010). Interestingly, reelin may interact with Notch signaling to control neuronal migration (Hashimoto-Torii et al. 2008). In addition, frizzled-related protein is a direct target of TH in the gut and may mediate TH actions on cell proliferation through the Wnt/β-catenin pathway (Kress et al. 2009). Although Wnt signaling is known to play an important role in development of the cortex (Solberg et al. 2008), it is not clear that TH plays a role in it. Taken together, these studies are identifying the direct gene targets of TH action and the transcriptional profile of the brain that is guided by TH. However, they also highlight a number of variables that will be important to control and study, including the timing of manipulation and evaluation, the region of the brain under consideration, and the pattern of thyroid hormone replacement when employed. In this regard, Royland and colleagues (2008) developed a unique and important strategy, employing different dose levels of PTU to manipulate thyroid status during development and identifying different dose–response relationships between circulating levels of TH and gene expression in both the cortex and hippocampus using microarray analysis.

Direct targets of TH action in the developing cerebellum have been identified using chromatin immunoprecipitation (Dong et al. 2009). This study identified a number of new targets of TH action, and this strategy is likely to provide important new information about the direct targets of TH action. In this context, it is important to consider the positions within the genome that TRs were found associated with DNA (Fig. 1). Specifically, the preponderance of TR binding sites was observed in

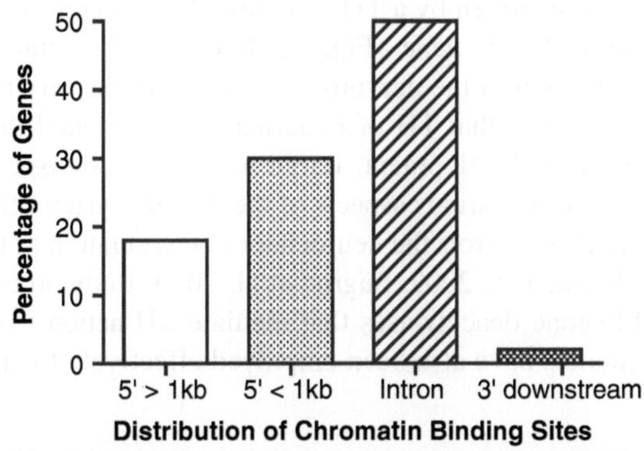

Fig. 1 Distribution of genomic locations of 91 TR binding sites confirmed by focused ChIP. Interestingly, 50% of all confirmed binding sites were present in introns. This may be an important theme for the way in which thyroid hormone receptors regulate gene expression. Fewer sites were observed in the region greater than 1 kb upstream from known promoter regions. This may also be an important motif for thyroid hormone receptors (Redrawn with permission from Dong et al. 2009)

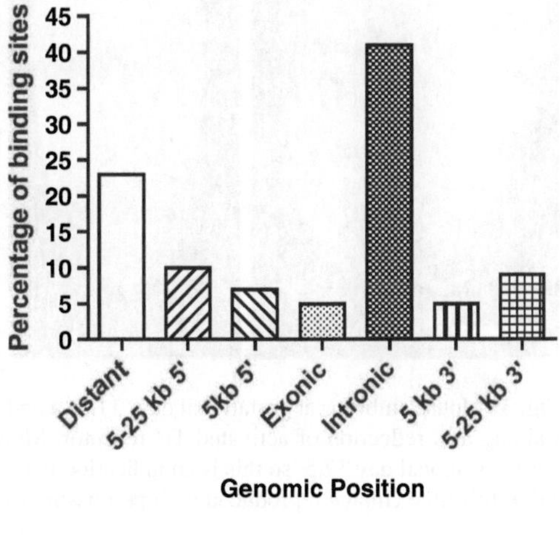

Fig. 2 Distribution of genomic locations of ERα binding sites in MCF-7 cells. Interestingly, 50% of all confirmed binding sites were present in introns. This may be an important theme for hormone-dependent nuclear receptors (Redrawn with permission from Welboren et al. 2009)

intronic sequences, which is consistent with ChIP studies for the estrogen receptor (Fig. 2), but a considerable number were also found in areas of chromatin far distant from known transcription start sites (Welboren et al. 2009). In addition, this strategy of identifying direct targets of TR binding to DNA is likely to give us a clearer idea of the specific developmental events affected by TH during development, as well as provide important information about the DNA regulatory elements to which the TRs are directed. For example, TH can regulate the expression of a microRNA in liver involved in the regulation of separate mRNAs (Dong et al. 2010), providing a potentially important glimpse at the variety of nuclear mechanisms employed by TH to regulate gene expression.

Genomic strategies for identifying the role of TH in brain development can be further enhanced by the development of transgenic approaches. For example, Nucera et al. (2010) recently engineered a mouse line expressing a transgene in which β-galactosidase is driven by a TH response element (TRE) derived from the myelin basic protein (MBP) gene (Fig. 3). Interestingly, fetal mice exhibited β-galactosidase expression in the nervous system before the onset of fetal thyroid function, indicating clearly that TH of maternal origin reached the fetus to drive gene expression early in development. Large numbers of transgenic mouse lines have been created to probe various aspects of the thyroid system, from elements of the feedback system that controls the neuroendocrine regulation of thyroid function (Chiamolera and Wondisford 2009; Sugrue et al. 2010) to the interaction between co-regulators and histone deacetylases that mediate TH action (You et al. 2010). These transgenic models have also been employed effectively to clarify the role of

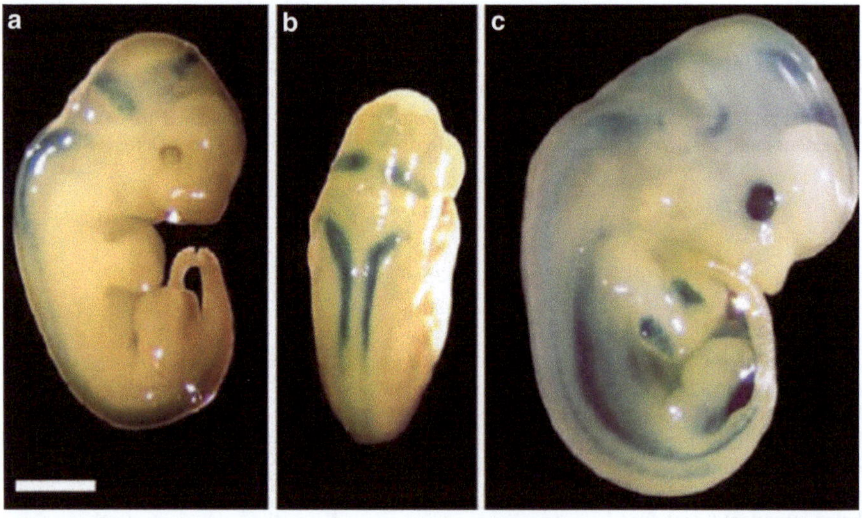

Fig. 3 Mouse embryos at gestational days 11.5 (**a** and **b**) and (**c**) 12.5 showing beta-galactosidase staining as a reflection of activated TH receptor. Mouse embryos do not produce their own TH until gestational day 17.5, so this is an indication that TH from the maternal circulation drives the TR at this time (Image reproduced with permission from Nucera et al. 2010)

TH and of TH metabolism in the development of the cochlea and retina (Jones et al. 2007; Ng et al. 2009; Pessoa et al. 2008; Roberts et al. 2006; Trimarchi et al. 2008). The continued use of genomics and genetics will be instrumental in discerning the complex actions of TH on brain development.

Importance of TH Delivery and Metabolism to Brain Development

There are a number of important processes that control the availability of T_3 to cells that may be acting independently of circulating levels of TH. TH circulates in the blood in rapid equilibrium with a variety of serum binding proteins, including Thyroxine Binding Protein (TBP), transthyretin (TTR) and albumin (Schussler 2000), and changes in the abundance of these proteins – or in T_4 binding to these proteins – may (Smallridge et al. 2005) or may not (Schussler 2000) be associated with physiological impact. In addition, both T_3 and T_4 can be actively transported into cells (Visser et al. 2011), and changes in the expression or function of specific transporters may have important developmental consequences (e.g., Heuer and Visser 2009; Schweizer et al. 2008). Finally, T_4 is "activated" to T_3 locally by type 1 (Dio1) or type 2 (Dio2) deiodinase, and it is "inactivated" to T2 or reverse T_3 by the type 3 deiodinase (Dio3; Schweizer et al. 2008). Interestingly, T_4 is converted to T_3 by Dio2 in glial cells in the brain and this T_3 is then delivered to neurons, where it is then metabolized by Dio3 (Bernal 2005). This spatial separation of deiodinase expression in the brain is likely to be extremely important.

Importantly, in studies to characterize targeted deletion of specific deiodinase enzymes, D3 knock-out mice exhibited the most severe neurodevelopmental phenotype among the three types (Galton et al. 2007; Hernandez et al. 2007, 2010), though this finding does not always coincide with deiodinase defects in humans (Schomburg 2010). In addition to metabolism, T_4 and T_3 transporters are likely to figure prominently in the normal regulation of TH action during development (Visser et al. 2010) and may well be important targets of endocrine disruptors (Westholm et al. 2009).

Given the importance of T_3 delivery to target cells for normal brain development, it is reasonable to propose that changes in the expression or function of proteins involved in this delivery process could represent an effective adaptive response that could compensate for changes in serum levels of TH during development. Although this is a logical hypothesis, there is little formal support for it. The observation that small reductions in serum T_4 alter the balance of production of oligodendrocytes and astrocytes during brain development (Sharlin et al. 2008) is not consistent with a robust compensatory apparatus. In addition, adaptive changes in serum TSH, in Dio2 expression and activity, and in MCT8 transporter expression were not associated with a change in the relationship between serum T_4 and the expression of a direct target of TH action (Sharlin et al. 2010). In humans, too, marginal iodine insufficiency (Aghini Lombardi et al. 1995; Berbel et al. 2009; Choudhury and Gorman 2003; Melse-Boonstra and Jaiswal 2010), maternal hypothyroxinemia and subclinical hypothyroidism (Haddow et al. 1999; Pop et al. 2003; Pop and Vulsma 2005) are associated with measurable cognitive deficits. Thus, the ability of the developing brain to compensate for long-term but small reductions in serum TH in both humans and in animals may not be as great as is often assumed.

Taken together, these studies are beginning to reveal the mechanisms regulating TH delivery to and action within developing tissues, as well as the developmental events under control of TH. They will undoubtedly clarify the role of TH in development and will ultimately guide clinical practice. It will be important to continue to develop and employ transgenic mouse models that allow us to better understand the causal relationships between TH and brain development. An important example is that of Flamant and Quignodon (2010), who described a new method of controlling the expression of a TR isoform. These studies also identify the many points of regulation of TH action that can be targeted by environmental chemicals. There are a very large number of manmade chemicals to which the human population is exposed that can affect circulating levels of THs (Brucker-Davis 1998; Howdeshell 2002). However, it is becoming clearer that environmental chemicals can affect TH action during development by many mechanisms, perhaps without influencing circulating levels of TH (Zoeller 2003). Examples are highlighted below to illustrate these points.

Polychlorinated Biphenyls (PCBs)

PCBs are a family of 209 congeners made up of two linked phenyl rings and varying degrees of chlorination (Erickson 2001). These chemicals are contaminants routinely found in the environment and in humans, despite their production being banned in the late 1970s. Because of their environmental persistence and their bioaccumulation, they are uniformly found in human tissues, including amniotic fluid, cord blood and breast milk (Barr et al. 2007). A number of studies have identified an association between PCB body burden and measures of cognitive deficit in children (Schantz et al. 2003). Stewart et al. (2003a, b, 2005, 2008) found a significant association between PCB concentration in cord blood and measures of cognitive function that was stronger in children with a smaller sub-region of the corpus callosum.

These and other studies demonstrate that background levels of PCB exposure can produce effects on brain development and that some of these effects are similar to those observed in children with thyroid dysfunction. Thus, an important theory that has been proposed to explain at least in part the neurotoxic effects of PCBs in the human population is that PCBs can interfere with TH signaling. This hypothesis has strong support in animal models but limited support in humans. However, experimental studies indicate that PCB exposure has complex actions on TH signaling and that, therefore, the studies designed to test this hypothesis in humans must address these complexities. This issue is briefly reviewed here.

First, PCB exposure in animal models nearly uniformly causes a reduction in circulating TH levels (Zoeller 2010). This finding is true for a wide variety of individual PCB congeners (Martin and Klaassen 2010), although it is certainly related to the dose. PCB exposure causes a reduction in serum total and free T_4 (e.g., Bansal and Zoeller 2008) as well as serum T_3 (e.g., Goldey et al. 1995). PCB exposure can also influence various measures of behaviors in rodent systems (see Roegge and Schantz 2006), but it is not always possible to ascribe these effects to a reduction in serum TH because these studies did not always employ TH replacement in their design. A number of studies have shown that PCB exposure can affect various neuroananatomical features of the brain and that at least some of these are similar to those observed in animals in which serum TH was perturbed (see Roegge et al. 2006). Perhaps the best example of PCB exposure producing effects on brain development by interfering with TH signaling was developed over an extensive series of studies by Crofton (2004) on hearing. Thus, in animals, there is strong support for the conclusion that PCB exposure can cause a reduction in serum TH levels, and there is strong support only for the case of hearing development that this causes adverse effects on brain development.

As might be predicted, the evidence that PCB exposure causes a reduction in serum TH levels is mixed in humans, as has been extensively reviewed recently (Jugan et al. 2010). Some reviewers conclude that support for the hypothesis that background levels of PCBs (focusing on dioxin-like compounds) can affect serum TH levels in humans is weak (Goodman et al. 2010), whereas others conclude that the evidence is moderately strong but that confounding variables make this a very difficult issue to resolve (Langer et al. 2006). Two specific points are noteworthy. First, different PCB congeners – or a combination of PCB congeners – may be more-or-less potent at affecting serum thyroid hormones, and using a summary statistic for exposure may be misleading (Yang et al. 2010). Second, there are biological variables, such as birth mode, that can influence serum TH levels in a manner that could obscure observations of interest if not considered in the analysis (Herbstman et al. 2008a, b).

Studies in animals and in vitro also indicate that PCBs may be affecting TH signaling independent of serum hormone levels, and this is an issue that will be important, though difficult, to address in humans. Our early work shows that a mixture of PCBs (A1254) produces a "TH-like" effect in the brain despite causing a reduction in serum TH (Zoeller et al. 2000). We followed this work with a number of studies that, in part, characterized the impact of PCB exposure on TH endpoints in the developing brain and showed that the effects were not the same as hypothyroidism itself (Bansal et al. 2005; Gauger et al. 2004; Sharlin et al. 2006). In fact, some endpoints responded to PCB exposure in a manner that was consistent with hypothyroidism, some responded in a manner consistent with hyperthyroidism (Bansal et al. 2005), and some did not respond (Bansal and Zoeller 2008; Fig. 4). This observation is also very similar to that of Roegge et al. (2006), who reported effects of PCB exposure on cerebellar Purkinje cells.

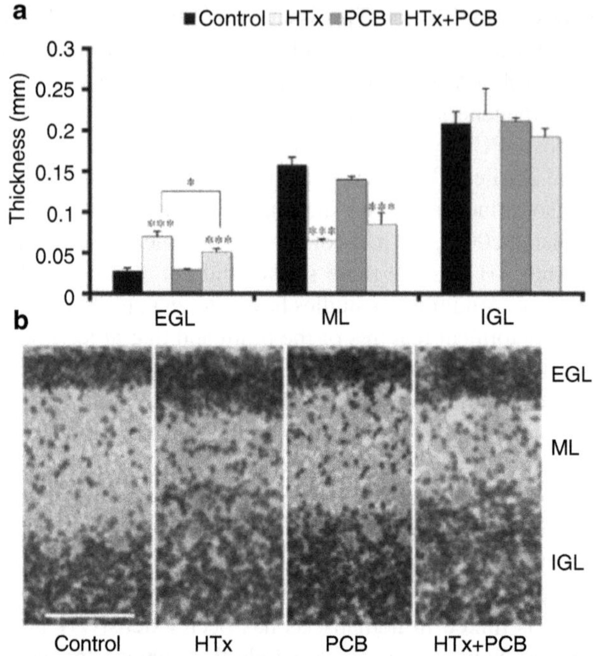

Fig. 4 Effect of PCB exposure in the presence or absence of induced hypothyroidism on cerebellar histogenesis. (**a**) Graphical representation of cerebellar layer thickness illustrated in (**b**). Although PCB exposure had no direct effect, it reduced the ability of goitrogen treatment (methimazole) to alter cerebellar development (Reproduced with permission from Bansal and Zoeller 2008)

We and others have explored the possibility that some PCB congeners or metabolites can act as direct agonists on the TR (You et al. 2006). Importantly, we found that a mixture of PCB congeners is required to activate transcription through the TR in vitro (Gauger et al. 2007). These data indicate that some PCB congeners can activate the expression of specific metabolic enzymes, which then hydroxylate other congeners that can act directly on the TR. These data are particularly noteworthy because they potentially explain why hydroxylated PCB metabolites are most strongly associated with cognitive deficits (Park et al. 2009). That is, perhaps specific non dioxin-like PCBs are metabolized (hydroxylated) to form compounds that inappropriately interact with the TR and maybe this interaction is responsible in part for changes in brain development that lead to measurable deficits in cognitive function. If this is true, then epidemiological studies of PCB body burden should examine the relationship between the combination of specific non-coplanar PCBs and the total amount of dioxin-like activity ("TEQ"). Thus, neither TEQ itself nor the abundance of specific non-coplanar PCBs would be associated with measures of cognitive function, but the combination would be a predictor of cognitive deficits.

It is also important to recognize that an early speculation may be correct: i.e., PCBs may act as imperfect ligands on the TR and their effects on the receptor may not be consistent with those of TH itself (McKinney and Waller 1998). For example, Koibuchi and his colleagues showed that specific PCBs could interfere with TR binding to a TH response element, which might explain the observation that some PCBs can interfere with TH action on primary cerebellar granule cells in culture (Amano et al. 2010; Miyazaki et al. 2004, 2008; Okada et al. 2005). Taken together, these studies show that background exposure to PCBs can influence brain development, and separate studies are variable in showing a relationship with measures of thyroid function. However, experimental studies reveal that the effect of PCB exposure on TH signaling is highly complex and this complexity has not been incorporated into studies in humans.

There are many biphenolic compounds to which the human population is exposed routinely, and early findings indicate that these compounds may also interact in similarly complex ways with the TH signaling system. These include flame retardants such as polybrominated diphenyls (PBDEs), halogenated bisphenol A (or parent BPA), polybrominated biphenyls, triclosan and triclocarban, and perhaps others. A potentially important example is illustrated by Decherf et al. (2010), who demonstrated that tetrabromobisphenol A (TBBPA) could influence the expression of the neuropeptide responsible for establishing the set-point around which the hypothalamic-pituitary-thyroid axis is controlled, but that the direction of influence depended upon the timing of TBBPA exposure. Thus, a number of these compounds may interfere with the TH system in ways that are not easily predicted or measured in humans.

Suvorov and Takser (2008) developed a very thoughtful case that, 50 years after PCB production was banned, the ways in which PBDEs and other biphenolic compounds are being studied appear to ignore what we learned from PCBs. This failure not only leads to the production and release of chemicals that have not been appropriately evaluated for safety but also leads to an inefficient exploration of toxicity once human populations are exposed. Overall, these studies demonstrate that environmental chemicals known to be associated with cognitive deficits in humans can interfere with TH signaling in ways that are difficult to observe and fully understand even in controlled animal studies, and in humans these studies will be difficult to pursue indeed.

Perchlorate

Environmental perchlorate contamination represents a case of environmental impacts on the thyroid with potentially important lessons. This issue has been reviewed extensively recently (Zewdie et al. 2010) and I will not provide a comprehensive description of the data here. Rather, I will highlight specific elements of the data to illustrate what may be important characteristics of thyroid endocrinology that present challenges for public health protection.

Perchlorate is used extensively as an oxidant in solid rocket fuels, fireworks, explosives, and other minor applications (Urbansky 2002). Because of its chemical stability in the environment, it has become widespread in the environment, contaminating water supplies and a wide range of foods (Murray et al. 2008). In turn, the human population has become widely contaminated, with perchlorate being measured in adult urine (Blount et al. 2006, 2007), amniotic fluid (Blount et al. 2009; Blount and Valentin-Blasini 2006) and breast milk (Kirk et al. 2007; Pearce et al. 2007). This exposure profile is of concern because perchlorate is a potent inhibitor of the sodium/iodide symporter (NIS) that is responsible for concentrating iodine in the thyroid gland, which is required for TH synthesis. Thus, environmental perchlorate could inhibit iodine uptake to such an extent that serum TH levels would decline.

To address this concern, several investigators explored the relationship between perchlorate exposure and iodide uptake inhibition in humans (Greer et al. 2002; Lawrence et al. 2000, 2001). These studies indicated that drinking water concentrations of about 200 μg/L would be required to begin to cause a measurable reduction in iodide uptake. The NRC Committee to Assess the Health Implications of Perchlorate Ingestion estimated that it would require a significantly higher concentration of perchlorate ingestion for a period of months to be able to impact circulating levels of TH (NRC 2005). This estimate was criticized by some because it focused on the adult, not the infant, which has significantly different thyroid characteristics (Ginsberg et al. 2007). Specifically, the adult thyroid gland has potentially several months of hormone stored in its colloid, but a newborn has none. In addition, the half-life of T_4 in serum is 7–10 days in the adult and about 3 days in the newborn (van den Hove et al. 1999). These differences were predicted to have important implications for the sensitivity of infants to perchlorate.

Considering these findings, it was surprising that Blount et al. (2006) found an association between urinary perchlorate and measures of thyroid function in U.S. women. Urinary perchlorate was a significant predictor of both T_4 and TSH in all women and this association was greater in women with lower levels of urinary iodine. Moreover, using the same data set, Steinmaus et al. (2007) showed that the relationship between perchlorate and measures of thyroid function were stronger in women who smoked. Because thiocyanate from cigarettes also competes with iodine at the NIS, these data are concordant with the hypothesis that background levels of perchlorate are affecting thyroid function. Finally, Cao et al. (2010) more recently found a similar association between perchlorate exposure and measures of thyroid function in infants.

Although it is premature at this time to conclude that environmental perchlorate contamination is causing adverse effects in population health, it is important to recognize that our understanding of the thyroid system, as articulated in the NRC report on perchlorate (NRC 2005), did not lead us to predict that long-term, low-dose exposure to perchlorate could influence measures of thyroid function in adults. It seems that there are two general possibilities to account for this enigma. One is that the biomonitoring studies of Blount et al. (2006, 2007) do not represent a causal relationship between perchlorate exposure and thyroid function, and our general understanding of the thyroid system is correct. Certainly, these studies need to be confirmed, but there does not appear to be a fundamental design flaw. The other possibility is that the short-term (2 weeks), high-dose studies of Greer et al. (2002) do not provide physiologically relevant information to extrapolate directly to lifetime low-dose exposures. This latter possibility requires some consideration.

A prevailing perspective about the thyroid system is that it is both driven by and contributes to homeostasis; as a result, there are compensatory mechanisms in place to both ameliorate the consequences of perturbations to the thyroid system and to ameliorate the consequences of thyroid perturbations downstream of TH action. As an index of the degree to which this concept pervades our thinking, it was invoked over 30 times in the NRC document on perchlorate (NRC 2005), but there were no references to scientific work that would support the way the concept was being employed. Thus, we make assumptions about the resilience of the thyroid system to chemical perturbation as well as the resilience of physiological systems to changes in thyroid function or TH action.

Our lab recently tested the hypothesis that a variety of potentially adaptive responses in the developing rat brain to low serum TH could ameliorate (or "compensate") for this perturbation and we found no evidence that this was the case (Sharlin et al. 2010). However, these studies were performed under conditions of chronic exposure to the goitrogenic agent PTU, and the compensatory responses of the thyroid system may be different under conditions of long-term and short-term perturbation. Minimally, we need to define the terms "homeostasis" and "compensation" in very precise terms and ensure that there is scientific support for our statements. This situation may be similar to the debate in hypertension research

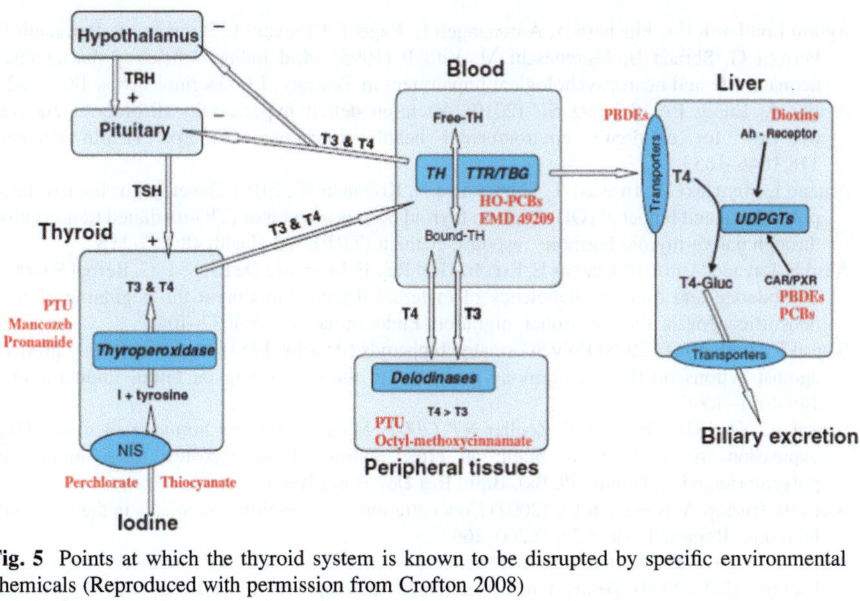

Fig. 5 Points at which the thyroid system is known to be disrupted by specific environmental chemicals (Reproduced with permission from Crofton 2008)

about the role of the baroreceptor reflex in the long-term and short-term control of blood pressure (Thrasher 2006).

Conclusion

The human population is exposed to a large number of industrial contaminants that are known to interfere with thyroid function or with TH action. Few studies have explored the ways in which these contaminants can interact with the thyroid system, but this may be an important avenue for research to provide insight into public health risks to these exposures (Fig. 5). Mechanistic studies are revealing complex interactions of certain classes of chemical contaminants on the TR, and while the number of contaminants that exert this action may be relatively restricted (Kohrle 2008; Schmutzler et al. 2007), they are uniformly present in the human population. Moreover, studies from a variety of domains suggest that we have an incomplete understanding of the thyroid system that is required to make educated guesses about the relative risk to the human population to exposures to these EDCs.

Acknowledgments Work reported in this manuscript from the author's laboratory was supported in part by NIH Grant ES010026 (RTZ), U.S. EPA Grants RD3213701 (RTZ) and R832134 (DSS), and from a generous gift from Mrs. Audrey McMahon.

References

Aghini Lombardi FA, Pinchera A, Antonangeli L, Rago T, Chiovato L, Bargagna S, Bertucelli B, Ferretti G, Sbrana B, Marcheschi M, Vitti P (1995) Mild iodine deficiency during fetal/neonatal life and neuropsychological impairment in Tuscany. J Endocrinol Invest 18:57–62

Aguiar A, Eubig PA, Schantz SL (2010) Attention deficit hyperactivity disorder: a focused overview for children's environmental health researchers. Environ Health Perspect 118:1646–1653

Amano I, Miyazaki W, Iwasaki T, Shimokawa N, Koibuchi N (2010) The effect of hydroxylated polychlorinated biphenyl (OH-PCB) on thyroid hormone receptor (TR)-mediated transcription through native-thyroid hormone response element (TRE). Ind Health 48:115–118

Auso E, Lavado-Autric R, Cuevas E, Escobar Del Rey F, Morreale De Escobar G, Berbel P (2004) A moderate and transient deficiency of maternal thyroid function at the beginning of fetal neocorticogenesis alters neuronal migration. Endocrinology 145:4037–4047

Bansal R, Zoeller RT (2008) Polychlorinated biphenyls (Aroclor 1254) do not uniformly produce agonist actions on thyroid hormone responses in the developing rat brain. Endocrinology 149:4001–4008

Bansal R, You SH, Herzig CT, Zoeller RT (2005) Maternal thyroid hormone increases HES expression in the fetal rat brain: an effect mimicked by exposure to a mixture of polychlorinated biphenyls (PCBs). Brain Res Dev Brain Res 156:13–22

Barr DB, Bishop A, Needham LL (2007) Concentrations of xenobiotic chemicals in the maternal-fetal unit. Reprod Toxicol 23(3):260–266

Berbel P, Mestre JL, Santamaria A, Palazon I, Franco A, Graells M, Gonzalez-Torga A, de Escobar GM (2009) Delayed neurobehavioral development in children born to pregnant women with mild hypothyroxinemia during the first month of gestation: the importance of early iodine supplementation. Thyroid 19:511–519

Bernal J (2005) The significance of thyroid hormone transporters in the brain. Endocrinology 146:1698–1700

Billon N, Tokumoto Y, Forrest D, Raff M (2001) Role of thyroid hormone receptors in timing oligodendrocyte differentiation. Dev Biol 235(1):110–120

Billon N, Jolicoeur C, Tokumoto Y, Vennstrom B, Raff M (2002) Normal timing of oligodendrocyte development depends on thyroid hormone receptor alpha 1 (TRalpha1). EMBO J 21:6452–6460

Blasi V, Longaretti R, Giovanettoni C, Baldoli C, Pontesilli S, Vigone C, Saccuman C, Nigro F, Chiumello G, Scotti G, Weber G (2009) Decreased parietal cortex activity during mental rotation in children with congenital hypothyroidism. Neuroendocrinology 89:56–65

Bloom B, Cohen RA, Freeman G (2009) Summary health statistics for U.S. children: National Health Interview Survey, 2007. Vital Health Stat 10:1–80

Blount BC, Valentin-Blasini L (2006) Analysis of perchlorate, thiocyanate, nitrate and iodide in human amniotic fluid using ion chromatography and electrospray tandem mass spectrometry. Anal Chim Acta 567:87–93

Blount BC, Pirkle JL, Osterloh JD, Valentin-Blasini L, Caldwell KL (2006) Urinary perchlorate and thyroid hormone levels in adolescent and adult men and women living in the United States. Environ Health Perspect 114:1865–1871

Blount BC, Valentin-Blasini L, Osterloh JD, Mauldin JP, Pirkle JL (2007) Perchlorate exposure of the US population, 2001–2002. J Expo Sci Environ Epidemiol 17:400–407

Blount BC, Rich DQ, Valentin-Blasini L, Lashley S, Ananth CV, Murphy E, Smulian JC, Spain BJ, Barr DB, Ledoux T, Hore P, Robson M (2009) Perinatal exposure to perchlorate, thiocyanate, and nitrate in New Jersey mothers and newborns. Environ Sci Technol 43:7543–7549

Brucker-Davis F (1998) Effects of environmental synthetic chemicals on thyroid function. Thyroid 8:827–856

Cao Y, Blount BC, Valentin-Blasini L, Bernbaum JC, Phillips TM, Rogan WJ (2010) Goitrogenic anions, thyroid-stimulating hormone, and thyroid hormone in infants. Environ Health Perspect 118:1332–1337

Chiamolera MI, Wondisford FE (2009) Minireview: thyrotropin-releasing hormone and the thyroid hormone feedback mechanism. Endocrinology 150:1091–1096

Choudhury N, Gorman KS (2003) Subclinical prenatal iodine deficiency negatively affects infant development in Northern China. J Nutr 133:3162–3165

Corbetta C, Weber G, Cortinovis F, Calebiro D, Passoni A, Vigone MC, Beck-Peccoz P, Chiumello G, Persani L (2009) A 7-year experience with low blood TSH cutoff levels for neonatal screening reveals an unsuspected frequency of congenital hypothyroidism (CH). Clin Endocrinol (Oxford) 71:739–745

Crofton KM (2004) Developmental disruption of thyroid hormone: correlations with hearing dysfunction in rats. Risk Anal 24:1665–1671

Crofton KM (2008) Thyroid disrupting chemicals: mechanisms and mixtures. Int J Androl 31:209–223

Decherf S, Seugnet I, Fini JB, Clerget-Froidevaux MS, Demeneix BA (2010) Disruption of thyroid hormone-dependent hypothalamic set-points by environmental contaminants. Mol Cell Endocrinol 323:172–182

Diez D, Grijota-Martinez C, Agretti P, De Marco G, Tonacchera M, Pinchera A, de Escobar GM, Bernal J, Morte B (2008) Thyroid hormone action in the adult brain: gene expression profiling of the effects of single and multiple doses of triiodo-L-thyronine in the rat striatum. Endocrinology 149:3989–4000

Dimitropoulos A, Molinari L, Etter K, Torresani T, Lang-Muritano M, Jenni OG, Largo RH, Latal B (2009) Children with congenital hypothyroidism: long-term intellectual outcome after early high-dose treatment. Pediatr Res 65:242–248

Dong H, Wade M, Williams A, Lee A, Douglas GR, Yauk C (2005) Molecular insight into the effects of hypothyroidism on the developing cerebellum. Biochem Biophys Res Commun 330:1182–1193

Dong H, Yauk CL, Rowan-Carroll A, You SH, Zoeller RT, Lambert I, Wade MG (2009) Identification of thyroid hormone receptor binding sites and target genes using ChIP-on-chip in developing mouse cerebellum. PLoS One 4:e4610

Dong H, Paquette M, Williams A, Zoeller RT, Wade M, Yauk C (2010) Thyroid hormone may regulate mRNA abundance in liver by acting on microRNAs. PLoS One 5(8):e12136

Dowling AL, Zoeller RT (2000) Thyroid hormone of maternal origin regulates the expression of RC3/neurogranin mRNA in the fetal rat brain. Brain Res Mol Brain Res 82:126–132

Dowling ALS, Martz GU, Leonard JL, Zoeller RT (2000) Acute changes in maternal thyroid hormone induce rapid and transient changes in specific gene expression in fetal rat brain. J Neurosci 20:2255–2265

Dowling AL, Iannacone EA, Zoeller RT (2001) Maternal hypothyroidism selectively affects the expression of neuroendocrine-specific protein A messenger ribonucleic acid in the proliferative zone of the fetal rat brain cortex. Endocrinology 142:390–399

EPA US (1998) Guidelines for neurotoxicity risk assessment. National Technical Information Service Publication PB98-117831. http://www.epa.gov/ncea/pdfs/nurotox.pdf

Erickson MD (2001) PCB properties, uses, occurrence, and regulatory history. In: Robertson LW, Hansen LG (eds) PCBs: recent advances in environmental toxicology and health effects. The University Press of Kentucky, Lexington, pp xi–xxx

Eubig PA, Aguiar A, Schantz SL (2010) Lead and PCBs as risk factors for attention deficit hyperactivity disorder. Environ Health Perspect 118:1654–1667

Flamant F, Quignodon L (2010) Use of a new model of transgenic mice to clarify the respective functions of thyroid hormone receptors in vivo. Heart Fail Rev 15:117–120

Galton VA, Wood ET, St Germain EA, Withrow CA, Aldrich G, St Germain GM, Clark AS, St Germain DL (2007) Thyroid hormone homeostasis and action in the type 2 deiodinase-deficient rodent brain during development. Endocrinology 148:3080–3088

Gauger KJ, Kato Y, Haraguchi K, Lehmler HJ, Robertson LW, Bansal R, Zoeller RT (2004) Polychlorinated biphenyls (PCBs) exert thyroid hormone-like effects in the fetal rat brain but do not bind to thyroid hormone receptors. Environ Health Perspect 112:516–523

Gauger KJ, Giera S, Sharlin DS, Bansal R, Iannacone E, Zoeller RT (2007) Polychlorinated biphenyls 105 and 118 form thyroid hormone receptor agonists after cytochrome P4501A1 activation in rat pituitary GH3 cells. Environ Health Perspect 115:1623–1630

Ginsberg GL, Hattis DB, Zoeller RT, Rice DC (2007) Evaluation of the U.S. EPA/OSWER preliminary remediation goal for perchlorate in groundwater: focus on exposure to nursing infants. Environ Health Perspect 115:361–369

Goldey ES, Kehn LS, Lau C, Rehnberg GL, Crofton KM (1995) Developmental exposure to polychlorinated biphenyls (Aroclor 1254) reduces circulating thyroid hormone concentrations and causes hearing deficits in rats. Toxicol Appl Pharmacol 135:77–88

Goodman JE, Kerper LE, Boyce CP, Prueitt RL, Rhomberg LR (2010) Weight-of-evidence analysis of human exposures to dioxins and dioxin-like compounds and associations with thyroid hormone levels during early development. Regul Toxicol Pharmacol 58:79–99

Gothe S, Wang Z, Ng L, Kindblom JM, Barros AC, Ohlsson C, Vennstrom B, Forrest D (1999) Mice devoid of all known thyroid hormone receptors are viable but exhibit disorders of the pituitary-thyroid axis, growth, and bone maturation. Genes Dev 13:1329–1341

Greer MA, Goodman G, Pleus RC, Greer SE (2002) Health effects assessment for environmental perchlorate contamination: the dose response for inhibition of thyroidal radioiodine uptake in humans. Environ Health Perspect 110:927–937

Haddow JE, Palomaki GE, Allan WC, Williams JR, Knight GJ, Gagnon J, O'Heir CE, Mitchell ML, Hermos RJ, Waisbren SE, Faix JD, Klein RZ (1999) Maternal thyroid deficiency during pregnancy and subsequent neuropsychological development of the child. N Engl J Med 341:549–555

Harris KB, Pass KA (2007) Increase in congenital hypothyroidism in New York State and in the United States. Mol Genet Metab 91:268–277

Hashimoto K, Curty FH, Borges PP, Lee CE, Abel ED, Elmquist JK, Cohen RN, Wondisford FE (2001) An unliganded thyroid hormone receptor causes severe neurological dysfunction. Proc Natl Acad Sci USA 98:3998–4003

Hashimoto-Torii K, Torii M, Sarkisian MR, Bartley CM, Shen J, Radtke F, Gridley T, Sestan N, Rakic P (2008) Interaction between Reelin and Notch signaling regulates neuronal migration in the cerebral cortex. Neuron 60:273–284

Herbstman J, Apelberg BJ, Witter FR, Panny S, Goldman LR (2008a) Maternal, infant, and delivery factors associated with neonatal thyroid hormone status. Thyroid 18:67–76

Herbstman JB, Sjodin A, Apelberg BJ, Witter FR, Halden RU, Patterson DG, Panny SR, Needham LL, Goldman LR (2008b) Birth delivery mode modifies the associations between prenatal polychlorinated biphenyl (PCB) and polybrominated diphenyl ether (PBDE) and neonatal thyroid hormone levels. Environ Health Perspect 116:1376–1382

Hernandez A, Martinez ME, Liao XH, Van Sande J, Refetoff S, Galton VA, St Germain DL (2007) Type 3 deiodinase deficiency results in functional abnormalities at multiple levels of the thyroid axis. Endocrinology 148:5680–5687

Hernandez A, Quignodon L, Martinez ME, Flamant F, St Germain DL (2010) Type 3 deiodinase deficiency causes spatial and temporal alterations in brain T3 signaling that are dissociated from serum thyroid hormone levels. Endocrinology 151:5550–5558

Heuer H, Visser TJ (2009) Minireview: pathophysiological importance of thyroid hormone transporters. Endocrinology 150:1078–1083

Horn S, Heuer H (2010) Thyroid hormone action during brain development: more questions than answers. Mol Cell Endocrinol 315:19–26

Howdeshell KL (2002) A model of the development of the brain as a construct of the thyroid system. Environ Health Perspect 110(Suppl 3):337–348

Jones I, Ng L, Liu H, Forrest D (2007) An intron control region differentially regulates expression of thyroid hormone receptor {beta}2 in the cochlea, pituitary, and cone photoreceptors. Mol Endocrinol 21:1108–1119

Jugan ML, Levi Y, Blondeau JP (2010) Endocrine disruptors and thyroid hormone physiology. Biochem Pharmacol 79:939–947

Kemper AR, Ouyang L, Grosse SD (2010) Discontinuation of thyroid hormone treatment among children in the United States with congenital hypothyroidism: findings from health insurance claims data. BMC Pediatr 10:9

Kirk AB, Dyke JV, Martin CF, Dasgupta PK (2007) Temporal patterns in perchlorate, thiocyanate, and iodide excretion in human milk. Environ Health Perspect 115:182–186

Kobayashi K, Akune H, Sumida K, Saito K, Yoshioka T, Tsuji R (2009) Perinatal exposure to PTU decreases expression of Arc, Homer 1, Egr 1 and Kcna 1 in the rat cerebral cortex and hippocampus. Brain Res 1264:24–32

Kohrle J (2008) Environment and endocrinology: the case of thyroidology. Ann Endocrinol (Paris) 69:116–122

Korada M, Kibirige M, Turner S, Day J, Johnstone H, Cheetham T (2008) The implementation of revised guidelines and the performance of a screening programme for congenital hypothyroidism. J Med Screen 15:5–8

Kress E, Rezza A, Nadjar J, Samarut J, Plateroti M (2009) The frizzled-related sFRP2 gene is a target of thyroid hormone receptor alpha1 and activates beta-catenin signaling in mouse intestine. J Biol Chem 284:1234–1241

Kugelman A, Riskin A, Bader D, Koren I (2009) Pitfalls in screening programs for congenital hypothyroidism in premature newborns. Am J Perinatol 26(5):383–385

Langer P, Tajtakova M, Kocan A, Vlcek M, Petrik J, Chovancova J, Drobna B, Jursa S, Pavuk M, Trnovec T, Sebokova E, Klimes I (2006) Multiple organochlorine pollution and the thyroid. Endocr Regul 40:46–52

Lavado-Autric R, Auso E, Garcia-Velasco JV, Arufe Mdel C, Escobar del Rey F, Berbel P, Morreale de Escobar G (2003) Early maternal hypothyroxinemia alters histogenesis and cerebral cortex cytoarchitecture of the progeny. J Clin Invest 111:1073–1082
Lawrence JE, Lamm SH, Pino S, Richman K, Braverman LE (2000) The effect of short-term low-dose perchlorate on various aspects of thyroid function. Thyroid 10:659–663
Lawrence J, Lamm S, Braverman LE (2001) Low dose perchlorate (3 mg daily) and thyroid function. Thyroid 11:295
Martin L, Klaassen CD (2010) Differential effects of polychlorinated biphenyl congeners on serum thyroid hormone levels in rats. Toxicol Sci 117:36–44
McEwan IJ (2009) Nuclear receptors: one big family. Methods Mol Biol 505:3–18
McKinney JD, Waller CL (1998) Molecular determinants of hormone mimicry: halogenated aromatic hydrocarbon environmental agents. J Toxicol Environ Health B Crit Rev 1:27–58
Melse-Boonstra A, Jaiswal N (2010) Iodine deficiency in pregnancy, infancy and childhood and its consequences for brain development. Best Pract Res Clin Endocrinol Metab 24:29–38
Miyazaki W, Iwasaki T, Takeshita A, Kuroda Y, Koibuchi N (2004) Polychlorinated biphenyls suppress thyroid hormone receptor-mediated transcription through a novel mechanism. J Biol Chem 279:18195–18202
Miyazaki W, Iwasaki T, Takeshita A, Tohyama C, Koibuchi N (2008) Identification of the functional domain of thyroid hormone receptor responsible for polychlorinated biphenyl-mediated suppression of its action in vitro. Environ Health Perspect 116:1231–1236
Morte B, Manzano J, Scanlan T, Vennstrom B, Bernal J (2002) Deletion of the thyroid hormone receptor alpha 1 prevents the structural alterations of the cerebellum induced by hypothyroidism. Proc Natl Acad Sci USA 99:3985–3989
Morte B, Ceballos A, Diez D, Grijota-Martinez C, Dumitrescu AM, Di Cosmo C, Galton VA, Refetoff S, Bernal J (2010a) Thyroid hormone-regulated mouse cerebral cortex genes are differentially dependent on the source of the hormone: a study in monocarboxylate transporter-8- and deiodinase-2-deficient mice. Endocrinology 151:2381–2387
Morte B, Diez D, Auso E, Belinchon MM, Gil-Ibanez P, Grijota-Martinez C, Navarro D, de Escobar GM, Berbel P, Bernal J (2010b) Thyroid hormone regulation of gene expression in the developing rat fetal cerebral cortex: prominent role of the Ca^{2+}/calmodulin-dependent protein kinase IV pathway. Endocrinology 151:810–820
Murray CW, Egan SK, Kim H, Beru N, Bolger PM (2008) US Food and Drug Administration's total diet study: dietary intake of perchlorate and iodine. J Expo Sci Environ Epidemiol 18:571–580
Ng L, Hernandez A, He W, Ren T, Srinivas M, Ma M, Galton VA, St Germain DL, Forrest D (2009) A protective role for type 3 deiodinase, a thyroid hormone-inactivating enzyme, in cochlear development and auditory function. Endocrinology 150:1952–1960
NRC (2005) Health implications of perchlorate ingestion. National Academic Press, Washington DC
Nucera C, Muzzi P, Tiveron C, Farsetti A, La Regina F, Foglio B, Shih SC, Moretti F, Della Pietra L, Mancini F, Sacchi A, Trimarchi F, Vercelli A, Pontecorvi A (2010) Maternal thyroid hormones are transcriptionally active during embryo-foetal development: results from a novel transgenic mouse model. J Cell Mol Med 14:2417–2435
Oetting A, Yen PM (2007) New insights into thyroid hormone action. Best Pract Res Clin Endocrinol Metab 21:193–208
Okada J, Shimokawa N, Koibuchi N (2005) Polychlorinated biphenyl (PCB) alters acid-sensitivity of cultured neurons derived from the medulla oblongata. Int J Biochem Cell Biol 37:1368–1374
Park HY, Park JS, Sovcikova E, Kocan A, Linderholm L, Bergman A, Trnovec T, Hertz-Picciotto I (2009) Exposure to hydroxylated polychlorinated biphenyls (OH-PCBs) in the prenatal period and subsequent neurodevelopment in eastern Slovakia. Environ Health Perspect 117:1600–1606
Pathak A, Sinha RA, Mohan V, Mitra K, Godbole MM (2010) Maternal thyroid hormone before the onset of fetal thyroid function regulates reelin and downstream signaling cascade affecting neocortical neuronal migration. Cereb Cortex 21:11–21

Pearce EN, Leung AM, Blount BC, Bazrafshan HR, He X, Pino S, Valentin-Blasini L, Braverman LE (2007) Breast milk iodine and perchlorate concentrations in lactating Boston-area women. J Clin Endocrinol Metab 92:1673–1677

Pessoa CN, Santiago LA, Santiago DA, Machado DS, Rocha FA, Ventura DF, Hokoc JN, Pazos-Moura CC, Wondisford FE, Gardino PF, Ortiga-Carvalho TM (2008) Thyroid hormone action is required for normal cone opsin expression during mouse retinal development. Invest Ophthalmol Vis Sci 49:2039–2045

Polanczyk G, de Lima MS, Horta BL, Biederman J, Rohde LA (2007) The worldwide prevalence of ADHD: a systematic review and metaregression analysis. Am J Psychiatry 164:942–948

Pop VJ, Vulsma T (2005) Maternal hypothyroxinaemia during (early) gestation. Lancet 365:1604–1606

Pop VJ, Brouwers EP, Vader HL, Vulsma T, van Baar AL, de Vijlder JJ (2003) Maternal hypothyroxinaemia during early pregnancy and subsequent child development: a 3-year follow-up study. Clin Endocrinol (Oxford) 59:282–288

Portella AC, Carvalho F, Faustino L, Wondisford FE, Ortiga-Carvalho TM, Gomes FC (2010) Thyroid hormone receptor beta mutation causes severe impairment of cerebellar development. Mol Cell Neurosci 44:68–77

Rice DS, Curran T (2001) Role of the reelin signaling pathway in central nervous system development. Annu Rev Neurosci 24:1005–1039

Roberts MR, Srinivas M, Forrest D, Morreale de Escobar G, Reh TA (2006) Making the gradient: thyroid hormone regulates cone opsin expression in the developing mouse retina. Proc Natl Acad Sci USA 103:6218–6223

Roegge CS, Schantz SL (2006) Motor function following developmental exposure to PCBS and/or MEHG. Neurotoxicol Teratol 28:260–277

Roegge CS, Morris JR, Villareal S, Wang VC, Powers BE, Klintsova AY, Greenough WT, Pessah IN, Schantz SL (2006) Purkinje cell and cerebellar effects following developmental exposure to PCBs and/or MeHg. Neurotoxicol Teratol 28:74–85

Rovet J, Daneman D (2003) Congenital hypothyroidism: a review of current diagnostic and treatment practices in relation to neuropsychologic outcome. Paediatr Drugs 5:141–149

Royland JE, Parker JS, Gilbert ME (2008) A genomic analysis of subclinical hypothyroidism in hippocampus and neocortex of the developing rat brain. J Neuroendocrinol 20:1319–1338

Schantz SL, Widholm JJ, Rice DC (2003) Effects of PCB exposure on neuropsychological function in children. Environ Health Perspect 111:357–576

Schmutzler C, Gotthardt I, Hofmann PJ, Radovic B, Kovacs G, Stemmler L, Nobis I, Bacinski A, Mentrup B, Ambrugger P, Gruters A, Malendowicz LK, Christoffel J, Jarry H, Seidlova-Wuttke D, Wuttke W, Kohrle J (2007) Endocrine disruptors and the thyroid gland – a combined in vitro and in vivo analysis of potential new biomarkers. Environ Health Perspect 115(Suppl 1):77–83

Schomburg L (2010) Genetics and phenomics of selenoenzymes – how to identify an impaired biosynthesis? Mol Cell Endocrinol 322:114–124

Schussler GC (2000) The thyroxine-binding proteins. Thyroid 10:141–149

Schuurmans C, Guillemot F (2002) Molecular mechanisms underlying cell fate specification in the developing telencephalon. Curr Opin Neurobiol 12:26–34

Schweizer U, Weitzel JM, Schomburg L (2008) Think globally: act locally. New insights into the local regulation of thyroid hormone availability challenge long accepted dogmas. Mol Cell Endocrinol 289:1–9

Sharlin DS, Bansal R, Zoeller RT (2006) Polychlorinated biphenyls exert selective effects on cellular composition of white matter in a manner inconsistent with thyroid hormone insufficiency. Endocrinology 147:846–858

Sharlin DS, Tighe D, Gilbert ME, Zoeller RT (2008) The balance between oligodendrocyte and astrocyte production in major white matter tracts is linearly related to serum total thyroxine. Endocrinology 149:2527–2536

Sharlin DS, Gilbert ME, Taylor MA, Ferguson DC, Zoeller RT (2010) The nature of the compensatory response to low thyroid hormone in the developing brain. J Neuroendocrinol 22:153–165
Shibusawa N, Hollenberg AN, Wondisford FE (2003) Thyroid hormone receptor DNA binding is required for both positive and negative gene regulation. J Biol Chem 278:732–738
Smallridge RC, Glinoer D, Hollowell JG, Brent G (2005) Thyroid function inside and outside of pregnancy: what do we know and what don't we know? Thyroid 15:54–59
Soen Y, Mori A, Palmer TD, Brown PO (2006) Exploring the regulation of human neural precursor cell differentiation using arrays of signaling microenvironments. Mol Syst Biol 2:37
Solberg N, Machon O, Krauss S (2008) Effect of canonical Wnt inhibition in the neurogenic cortex, hippocampus, and premigratory dentate gyrus progenitor pool. Dev Dyn 237:1799–1811
Steinmaus C, Miller MD, Howd R (2007) Impact of smoking and thiocyanate on perchlorate and thyroid hormone associations in the 2001–2002 national health and nutrition examination survey. Environ Health Perspect 115:1333–1338
Stewart P, Fitzgerald S, Reihman J, Gump B, Lonky E, Darvill T, Pagano J, Hauser P (2003a) Prenatal PCB exposure, the corpus callosum, and response inhibition. Environ Health Perspect 111:1670–1677
Stewart PW, Reihman J, Lonky EI, Darvill TJ, Pagano J (2003b) Cognitive development in preschool children prenatally exposed to PCBs and MeHg. Neurotoxicol Teratol 25:11–22
Stewart P, Reihman J, Gump B, Lonky E, Darvill T, Pagano J (2005) Response inhibition at 8 and 9 1/2 years of age in children prenatally exposed to PCBs. Neurotoxicol Teratol 27:771–780
Stewart PW, Lonky E, Reihman J, Pagano J, Gump BB, Darvill T (2008) The relationship between prenatal PCB exposure and intelligence (IQ) in 9-year-old children. Environ Health Perspect 116:1416–1422
Sugrue ML, Vella KR, Morales C, Lopez ME, Hollenberg AN (2010) The thyrotropin-releasing hormone gene is regulated by thyroid hormone at the level of transcription in vivo. Endocrinology 151:793–801
Suvorov A, Takser L (2008) Facing the challenge of data transfer from animal models to humans: the case of persistent organohalogens. Environ Health 7:58
Takahashi M, Negishi T, Tashiro T (2008) Identification of genes mediating thyroid hormone action in the developing mouse cerebellum. J Neurochem 104:640–652
Thrasher TN (2006) Arterial baroreceptor input contributes to long-term control of blood pressure. Curr Hypertens Rep 8:249–254
Trimarchi JM, Harpavat S, Billings NA, Cepko CL (2008) Thyroid hormone components are expressed in three sequential waves during development of the chick retina. BMC Dev Biol 8:101
Urbansky ET (2002) Perchlorate as an environmental contaminant. Environ Sci Pollut Res Int 9:187–192
van den Hove MF, Beckers C, Devlieger H, de Zegher F, De Nayer P (1999) Hormone synthesis and storage in the thyroid of human preterm and term newborns: effect of thyroxine treatment. Biochimie 81:563–570
van der Sluijs VL, Kempers MJ, Last BF, Vulsma T, Grootenhuis MA (2008) Quality of life, developmental milestones, and self-esteem of young adults with congenital hypothyroidism diagnosed by neonatal screening. J Clin Endocrinol Metab 93:2654–2661
Visser WE, Friesema EC, Visser TJ (2011) Minireview: thyroid hormone transporters: the knowns and the unknowns. Mol Endocrinol 25:1–14
Welboren WJ, van Driel MA, Janssen-Megens EM, van Heeringen SJ, Sweep FC, Span PN, Stunnenberg HG (2009) ChIP-Seq of ERalpha and RNA polymerase II defines genes differentially responding to ligands. EMBO J 28:1418–1428
Weller M, Krautler N, Mantei N, Suter U, Taylor V (2006) Jagged1 ablation results in cerebellar granule cell migration defects and depletion of Bergmann glia. Dev Neurosci 28:70–80

Westholm DE, Stenehjem DD, Rumbley JN, Drewes LR, Anderson GW (2009) Competitive inhibition of organic anion transporting polypeptide 1c1-mediated thyroxine transport by the fenamate class of nonsteroidal antiinflammatory drugs. Endocrinology 150:1025–1032

Windham GC, Zhang L, Gunier R, Croen LA, Grether JK (2006) Autism spectrum disorders in relation to distribution of hazardous air pollutants in the san francisco bay area. Environ Health Perspect 114:1438–1444

Wondisford FE (2003) Thyroid hormone action: insight from transgenic mouse models. J Invest Med 51:215–220

Yang JM, Salmon AG, Marty MA (2010) Development of TEFs for PCB congeners by using an alternative biomarker–thyroid hormone levels. Regul Toxicol Pharmacol 56:225–236

You SH, Gauger KJ, Bansal R, Zoeller RT (2006) 4-Hydroxy-PCB106 acts as a direct thyroid hormone receptor agonist in rat GH3 cells. Mol Cell Endocrinol 257–258:26–34

You SH, Liao X, Weiss RE, Lazar MA (2010) The interaction between nuclear receptor corepressor and histone deacetylase 3 regulates both positive and negative thyroid hormone action in vivo. Mol Endocrinol 24:1359–1367

Zewdie T, Smith CM, Hutcheson M, West CR (2010) Basis of the Massachusetts reference dose and drinking water standard for perchlorate. Environ Health Perspect 118:42–48

Zoeller RT (2003) Thyroid toxicology and brain development: should we think differently? Environ Health Perspect 111:A628

Zoeller RT (2010) Environmental chemicals targeting thyroid. Hormones (Athens) 9:28–40

Zoeller RT, Rovet J (2004) Timing of thyroid hormone action in the developing brain: clinical observations and experimental findings. J Neuroendocrinol 16:809–818

Zoeller RT, Dowling AL, Vas AA (2000) Developmental exposure to polychlorinated biphenyls exerts thyroid hormone-like effects on the expression of RC3/neurogranin and myelin basic protein messenger ribonucleic acids in the developing rat brain. Endocrinology 141:181–189

Westholm DE, Stenehjem DD, Rumbley JN, Drewes LR, Anderson GW (2009) Competitive inhibition of organic anion transporting polypeptide 1c1-mediated thyroxine transport by the fenamate class of nonsteroidal antiinflammatory drugs. Endocrinology 150:1025–1032

[illegible]

Neural Progenitors Are Direct Targets of Xenoestrogens in Zebrafish

Olivier Kah, Yann Le Page, Mélanie Vosges, Sok-Keng Tong, Bon-chu Chung, and François Brion

Abstract Because a large proportion of endocrine disruptor chemicals (EDC) end up in surface waters, aquatic species are particularly vulnerable to their potential effects. In this regard, fish populations must be carefully monitored for fishes are absolutely crucial in terms of biodiversity and protein resources, but also they are extremely valuable as sentinel species. In this chapter, we discuss EDCs effects on the brain of fish, in particular on radial glial cells, which in all vertebrate species are brain stem cells. Indeed, one of the most prominent effect of EDCs in zebrafish is their impact on the *cyp19a1b* gene that encodes aromatase B. Strikingly, aromatase B is only expressed in radial glial cells that behave as neuronal progenitors. Detailed molecular and whole animal studies in transgenic zebrafish demonstrated the extreme sensitivity of the *cyp19a1b* gene to estrogen mimics. In particular, doses as low as 1.5 ng/L of EE2 were consistently shown to turn on *cyp19a1b* gene expression in 2–5 days old zebrafish embryos. As recent studies indicate that estrogens modulate proliferative activity of radial glia progenitors, it is likely that estrogen mimics may have similar activity. The potential outcome of such effects requires thorough investigations, not only in fish but also in developing mammals. In addition, those studies have led to the development of a very sensitive in vivo assay that makes use of *cyp19a1b*-GFP transgenic embryos whose brain exhibits GFP expression if exposed to any estrogen mimic acting through estrogen receptors.

Supported by the CNRS, the ANR NEED (CES 2008–011) and the Post-Grenelle program NEMO.

O. Kah (✉)
Neurogenesis and Estrogens, UMR CNRS 6026, IFR 140, Case 1302, The University of Rennes 1, Rennes cedex, France
e-mail: Olivier.kah@univ-rennes1.fr

J.-P. Bourguignon et al. (eds.), *Multi-System Endocrine Disruption*, Research and Perspectives in Endocrine Interactions,
DOI 10.1007/978-3-642-22775-2_5, © Springer-Verlag Berlin Heidelberg 2011

Introduction

It is now well established that a number of natural or man-made compounds, referred to as "Endocrine Disrupting Chemicals" (EDCs) interfere with the endocrine systems in wildlife, domestic animals and human beings. This is supported by an ever-increasing number of scientific publications showing a range of deleterious effects of these compounds many of which can be found in surface waters. Recent reports from different official bodies indicate that, in industrialized countries, it is now almost impossible to find surface waters free of contaminants (Allner et al. 2010). As a result, aquatic species, notably fishes, are primary targets of EDCs, and as it is the case in other vertebrates, EDC exposure is thought to be especially critical at the larval or developmental stages of fish, potentially causing decreased reproductive fitness in adults (Sharpe and Skakkebaek 2008).Teleost fish represent the largest group of actinopterygian fishes, a lineage that diverged some 450 millions years ago from the sarcopterygians fish, the tetrapod ancestors. This group experienced a rapid speciation and evolutive success, resulting in the existence of more than 30,000 species that are crucial in terms of biodiversity and protein resources. In some cases, viability of economically and ecologically relevant fish stocks can be at risk due to feminisation of male fish by a range of EDCs (Jobling et al. 2006; Kidd et al. 2007). Because fish biology and physiology are extensively studied, fishes are also valuable sentinel species that may be used to indicate contamination of surface waters with EDCs, provided that reliable, powerful and readily usable biomarkers are available enabling assessment of anthropogenic impacts.

Until now, most of the EDCs effects reported on fish have been concerned with peripheral functions such as vitellogenin synthesis in the liver (Vaillant et al. 1988; Flouriot et al. 1995; Sumpter and Jobling 1995; Petit et al. 1997) or gonadal sex change (Nagahama et al. 2004). In contrast only limited attention is dedicated to the central nervous system that is likely to be targeted by a wide range of EDCs. Over the last 5 years, our laboratories carried out intense research activities on estrogen signalling in the brain of zebrafish and we have identified several brain targets for EDCs. We have notably shown that EE2 disrupts the ontogeny of GnRH system by inducing a dose-dependent increase in the number of Gnrh3 neurons and a modification of the migration profile of Gnrh3 neurons. Such effects were observed at concentrations as low as 0.1 nM and were detectable in embryos from 5 days post-fecundation (dpf) to 20 dpf (Vosges et al. 2010). Studies by others have highlighted the effects of PCBs or fluoxetine on the monoaminergic systems of fish, causing impaired reproductive functions and disruption of feeding or sexual behavior (Khan and Thomas 2001, 2004; Thomas et al. 2007; Mennigen et al. 2009; Oakes et al. 2010).

In addition, accumulating data acknowledged aromatase B, the product of the *cyp19a1b* gene, as a highly sensitive target for estrogens and an outstanding biomarker for xenoestrogens exposure. As the *cyp19a1b* gene is strongly and only expressed in radial glial cells of developing and adult fish, these studies showed that estrogens and xenoestrogens directly target neuronal progenitors.

Indeed, as in all vertebrates, radial glial cells in fish are brain stem cells capable of generating neurons (Pellegrini et al. 2007; Kriegstein and Alvarez-Buylla 2009; Rowitch and Kriegstein 2010). Collectively, these data point to unique properties of the brain of fish with respect to brain development, sexualization and aromatase activity (Cheshenko et al. 2008; Mouriec et al. 2008; Diotel et al. 2010; Kah and Dufour 2010; Le Page et al. 2010). In particular, we have suggested that, to a certain extent, the brain of adult fish keeps properties of the developing brain in mammals (Diotel et al. 2010).

Aromatase Is Expressed in Radial Glial Cells of Adult Fish

Aromatase: A Key Enzyme that Regulates Estradiol Availability

In all vertebrates, cytochrome P450 aromatase (CYP19A1) plays a crucial role at all levels of the reproductive cycle by regulating the amounts of estrogens available for physiological effects. Aromatase is indeed the only enzyme capable of converting C19 androgens, such as testosterone or androstenedione, into C18 estrogens, estradiol or estrone, respectively. Initially believed to play key functions in females only, intensive research led to the discovery that aromatase and estrogens are also important in males notably for energy balance, bone formation, modulation of pituitary gonadotropins, spermatogenesis, and development of gynecomastia (Santen et al. 2009). Aromatase in mammals is expressed in multiple tissues through the use of alternative tissue-specific promoters. In addition to the gonads (both ovary and testis) aromatase in expressed in the adipocytes, skin fibroblasts, bone osteoblasts and osteoclasts, muscles, endothelial cells and the central nervous system (Simpson 2004; Boon et al. 2010). In all these tissues estrogens, notably estradiol (E2) exert multiple and powerful actions through a complex network of signalling pathways that includes both genomic and non-genomic mechanisms (Toran-Allerand 2004a).

The potential actions of E2 in the brain of vertebrates have received tremendous attention and this wealth of studies stressed out the importance of estrogens in regulating crucial aspects of brain development such as cell proliferation, apoptosis, cell migration, cell differentiation or synaptogenesis (McEwen 2001; Walf and Frye 2006; Sharpe and Skakkebaek 2008; Frye 2009). Historically, estrogens were first envisioned as “reproductive only” hormones and the focus was on regulation of neuroendocrine circuits and sexual behaviour. However, the recognized effects of estrogens in mammals are no longer limited to the “reproductive brain”, but now extend to modulation of cognitive functions, mood, motor control or neuroinflammation. In addition to the increasing complexity of estrogen signalling in target cells (Toran-Allerand 2004b), our understanding of the effects of estrogens in the brain is made more difficult by the fact that estrogens can reach the brain via the general circulation, or be the result of a local conversion of androgens by aromatase. In turn, this testosterone available for brain aromatization can be of peripheral origin or produced within the brain where many steroidogenic enzymes are expressed, including aromatase (Do Rego et al. 2009). It is believed that all three situations can occur, and even co-exist at a given time, making it rather difficult to extract a clear picture (Konkle and McCarthy 2011).

Aromatase Is Expressed in Radial Glial Cells of zebrafish

Intriguingly, aromatase expression is much higher in the brain of adult fish than in the brain of other adult vertebrates (Callard et al. 1978; Pasmanik and Callard 1985; Callard and Pasmanik 1987). It is also known that aromatase activity changes according to the reproductive season (Pasmanik and Callard 1985; Borg et al. 1989; Gonzalez and Piferrer 2003) suggesting a regulation by some gonadal factors. Recent studies have shown that aromatase activity in the brain of fish is due to the strong expression of *cyp19a1b*, one of two duplicated *cyp19a1* genes. This duplication is thought to have occurred early in the teleost lineage (Tchoudakova and Callard 1998; Diotel et al. 2010b). Interestingly, in situ hybridization and immunohistochemistry on different fish species revealed that *cyp19a1b* expression is confined to one type of brain cells with unique properties, the radial glial cells (RGC) (Forlano et al. 2001; Menuet et al. 2003, 2005; Strobl-Mazzulla et al. 2005; Marsh et al. 2006; Strobl-Mazzulla et al. 2008). RGCs have been first described by Magini at the end of the nineteenth century (Rakic 2003). These cells differentiate from neuroepithelial cells and are characterized by their typical morphology. They posses a small nucleus adjacent to the ventricular lumen, an apical process connecting the cerebrospinal fluid and a long basal process ending at the pial surface and characterized for their peculiar radial morphology (Cameron and Rakic 1991; Bentivoglio and Mazzarello 1999).

Historically, scientists thought that RGC had only a role of scaffold for neuronal migration (Rakic 1978; Bentivoglio and Mazzarello 1999), but it is now established that RGC are crucial for neurogenesis during embryonic development. Indeed, they are not only important for scaffolding, but in addition RGC are in fact progenitor cells giving rise to neurons (Noctor et al. 2001; Campbell and Gotz 2002).

Adult Neurogenesis in Fish Is Supported by Radial Glia Progenitors

Recent studies in the cortex of birds and mammals showed that RGCs, considered as belonging to the glial lineage, are in fact the neural stem cells that generate neurons and glial cells during development (Kriegstein and Alvarez-Buylla 2009; Costa et al. 2010; Taverna and Huttner 2010). During embryogenesis, such cells perform asymmetrical divisions and either give birth directly to neurons or to intermediate progenitor cells that in turn may produce neurons, oligodendrocytes or astrocytes (Jagasia et al. 2006; Ninkovic and Gotz 2007; Kriegstein and Alvarez-Buylla 2009). The role of estradiol in this process is suspected, but far from being clear (Martinez-Cerdeño et al. 2006). At the end of the developmental period, RGCs disappear from the mammalian brain. Their nucleus moves away from the ventricle, while the apical process loses its ventricular attachment. RGCs gain a multipolar astrocyte phenotype. In mammals, most RGCs transform into astrocytes while in birds, they persist in adult and continue to give birth to neurons in certain neurogenic hot spots (Kriegstein and Alvarez-Buylla 2009).

A well-recognized characteristic of the brains of fish is their capacity to constantly grow during "adulthood" as shown by pioneer studies in the guppy (Kirsche 1967; Kranz and Richter 1970a, 1970b). In contrast to mammals and birds in which adult neurogenesis is restricted to small brain regions (Lindsey and Tropepe 2006), these studies demonstrated extensive cell proliferation in the telencephalon, diencephalon, cerebellum and spinal cord of fish. Although the proliferative activity decreased progressively with age and size, the brain of the guppy still exhibits active proliferation in adult (Kranz and Richter 1970a). Since then, the existence of widespread periventricular proliferative zones (PZ) was confirmed in different teleost species (Ekstrom et al. 2001), notably in zebrafish (Zupanc et al. 2005; Adolf et al. 2006; Grandel et al. 2006; Pellegrini et al. 2007). This property is linked to the fact that fish keep growing during their entire lifespan, making it necessary to constantly generate new neurons (Brandstatter and Kotrschal 1989, 1990).

Very recent studies in zebrafish have now established that cell proliferation in the brain is associated with an important generation of new neurons, raising the question of the nature of the progenitors (Adolf et al. 2006; Pellegrini et al. 2007; Zupanc, 2001; Zupanc and Zupanc 2006). Using different markers of RGC (AroB, BLBP, GFAP, CXCR4, S-100, nestin) combined to either PCNA or BrdU immunohistochemistry, it has now been shown that most of the proliferative cells correspond to RGCs (Pellegrini et al. 2007; Diotel et al. 2010a; März et al. 2010). Thus, AroB-positive RGC have the capacity to divide and long term experiments using BrdU showed that newborn cells migrate along the radial processes in a way similar to what was described in the mouse cortex during embryogenesis (Pellegrini et al. 2007). Interestingly, newborn cells originating mainly from ventricular precursors migrate within the parenchyma to colonize different brain regions and to differentiate into neurons as shown by the use of markers such as HuC/D or acetylated tubulin (Adolf et al. 2006; Grandel et al. 2006; Pellegrini et al. 2007; Zupanc, 2006).

However, it appears that there are different subtypes of progenitor cells in the adult zebrafish telencephalon: (1) fast proliferating cells which are localized in the ventral subpallium and posterior pallium, (2) slow proliferating cells, corresponding to RGCs, in the other ventricular telencephalic regions (Adolf et al. 2006; Grandel et al. 2006) and (3) slow proliferating cells that do not correspond to RGCs (Adolf et al. 2006; März et al. 2010). This process is likely to support the constant growth of the brain of fish and somehow identifies the brain of adult fish as a "developing brain" in terms of neurogenesis (Fig. 1).

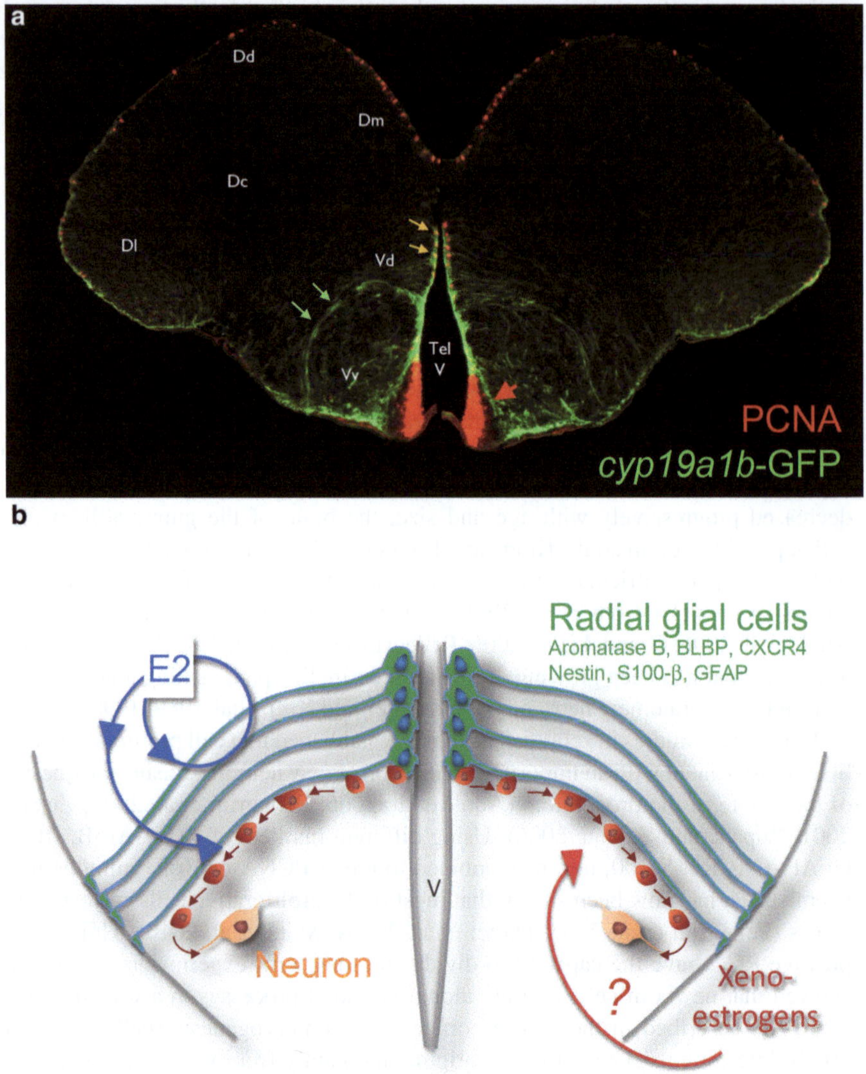

Fig. 1 (**a**) Transverse section through the telencephalon of a zebrafish showing aromatase B in the brain of a transgenic zebrafish expressing GFP under the control of the *cypl9alb* (aromatase B) promoter (Tong et al. 2009; Mouriec et al. 2010). GFP expression is restricted to radial glial cells and their long processes (*green arroHs*; Menuet et al. 2005; Pellegrini et al. 2007; Diotel et al. 2010). Many proliferative cells, stained in red by PCNA, are visible along the telencephalic ventricle. Some of these cells in yellow correspond to dividing RGCs performing asymmetrical divisions (*yelloH arroHs*; (März et al. 2010). The large *red arroH* points to a zone of rapidly cycling progenitors that give birth to neurons migrating towards the olfactory bulbs via the rostral migratory stream. Such cells do not express aromatase B (Pellegrini et al. 2007; Diotel et al. 2010a; März et al. 2010). (**b**) Glial nature of neuronal progenitors in development and in the adult brain of fish. Radial glial cells perform asymmetrical divisions and generate neurons that migrate along the radial processes before integrating functional circuits (Pellegrini et al. 2007). Radial glial cells express astroglial markers (Diotel et al. 2010a; März et al. 2010) together with aromatase B

Brain Aromatase in Fish Is Extremely Sensitive to Estrogens and Xenoestrogens

Many studies have documented the sensitivity of the *cyp19a1b* gene to estradiol (Callard et al. 2001; Cheshenko et al. 2008; Diotel et al. 2010b). In several species of fish, it is known that the promoter of the *cyp19a1b* gene exhibits a very well conserved ERE that is located at about −300 bp of the transcription start site. Transactivation experiments using a *cyp19a1b*-luciferase reporter gene have well documented that this ERE binds fish or human estrogen receptors with a high affinity and that its integrity is absolutely required for the E2 stimulation (Menuet et al. 2005; Mouriec et al. 2009). However, further studies have shown that a sequence called GxRE, located just upstream the ERE is also indispensable for full estrogenic response (Le Page et al. 2008). If mutated, or removed, the promoter looses around 80% of its responsiveness to E2. In gel shift studies, this sequence was shown to specifically binds extracts from cells of the astroglial lineage (Le Page et al. 2008). The hypothesis is thus that a mandatory cooperation exists between the estrogen receptors and an unknown glial factor that would confer a high E2 responsiveness to the *cyp19a1b* gene, explaining why this gene is expressed only in radial glial cells and not in the many neurons that express estrogen receptors in the brain of fish. Recent data in zebrafish have shown that indeed estrogen receptors are expressed in radial glial cells (I. Anglade et al. unpublished). The fact that aromatase expression is upregulated by estrogens implies that aromatizable androgens will also indirectly stimulate *cyp19a1b* expression as they are aromatized into estrogens in radial glial cells. This loop explains why aromatase expression is so strong in the brain of sexually mature fish with high levels of steroids (Callard et al. 1990). More recently, it was also demonstrated that some androgens (testosterone, dihydrotestosterone or 5α-androstane-3β,17β-diol) also have the property to up-regulate *cyp19a1b* gene expression in the brain of zebrafish, but these effects are mediated through estrogen receptors and not androgen receptors (Mouriec et al. 2009). Indeed, the brain of fish is also known for exhibiting a high expression of 5α-reductase (Pasmanik and Callard 1985) capable of converting testosterone into dihydrotestosterone that can be further metabolized into 5α-androstane-3β,17β-diol, an androgen with well known estrogenic properties (Mouriec et al. 2009). This substance (3Aβdiol), sometimes referred to as the "second estrogen" (Sugiyama et al. 2010), is produced by 17β-hydroxysteroid dehydrogenase type 7 (17βHSD VII) from dihydrotestosterone. This pathway is very active in the brain of fish as shown by the fact that DHY strongly up-regulates cyp19a1b expression in zebrafish. In zebrafish, 3Aβdiol binds all three estrogen receptors with a high affinity (Mouriec et al. 2009).

Fig. 1 (continued) (Menuet et al. 2005). These cells can produce estrogens that can act in an autocrine or paracrine manner to modulate the neurogenic process. Recent data indicate that estrogens modulate the proliferative activity of RGCs, an effect that can potentially be mimicked by xenoestrogens although this remains to be firmly established

Neural Progenitors Are Direct Targets for Xenoestrogens Endocrine Disruptors in Zebrafish

Given the fact that aromatase B is very sensitive to estrogens through ER dependent mechanisms, it is consequently a target for the many xenoestrogens and phytoestrogens that bind and activate ERs. In agreement, many reports indicate that a variety of xenoestrogens stimulate *cyp19a1b* mRNA expression, aromatase B protein expression or brain aromatase activity (Table 1). Our own studies in zebrafish confirmed the very high sensitivity of the *cypa19a1b* gene to xenoestrogens in vivo and in vitro (Le Page et al. 2006; Le Page et al. 2008; Mouriec et al. 2009; Vosges et al. 2010). Importantly, these studies showed that increased *cyp19a1b* expression is observed only in the radial cells, identifying RGCs as direct targets for xenoestrogens. The availability of transgenic zebrafish tg(*cyp19a1b-GFP)* expressing GFP under the control of the *cyp19a1b* promoter made it possible to clearly demonstrate that such cells are targeted by endocrine disruptors. Figure 2 shows that treating zebrafish embryos with recognized estrogens, aromatizable androgens, or dihydrotestosterone causes a dramatic expression of *cyp19a1b* only in the RGCs and never in neurons. Recent data showed that a range of xenoestrogens causes dose-dependent increase of GFP in the tg(*cyp19a1b-GFP*) zebrafish. This is notably the case of pharmaceuticals (hexestrol, diethystilbestrol, ethinylestradiol), phytoestrogens (genistein, α-zearalenol, α-zearalanol, β-zearalanol), nonylphenols (bisphenol A, 4-tpentylphenol, 4-t octylphenol) and the pesticides (methoxychlor, 2,2-bis(phydroxyphenyl)-1,1,1-trichloroethane; Brion et al. in preparation).

Concentrations of ethinylestradiol, as low as 5pM, are sufficient to up-regulate *cyp19a1B* expression in zebrafish embryos. This highlights both the great sensitivity of this gene and its potential as a biomarker of exposure. For example, Fig. 3 shows the dose dependent effects of 4-tert-octylphenol on *cyp19a1b*-GFP expression in the brain of a zebrafish embryo. Before being banned octylphenols were used in rubber, cleaning goods, pesticides and paints. They are recognized estrogenic chemicals (Hogan et al. 2006; Jespersen et al. 2010) and here we show that 4-tert-octylphenolat concentrations around 25–50 μg/L can enter the brain of fish and induce GFP expression in vivo.

Consequences of Xenoestrogens Actions on Progenitor Cells in Fish Brain

The above data clearly indicate that the brain of developing and adult fish is a target for the numerous xenoestrogens present in surface waters. What is unclear is to what extent such central actions of xenoestrogens contribute to affect the reproductive health. Indeed, the roles of aromatase B expression in radial progenitors are still unclear. It has been hypothesized that, in the presence of aromatizable androgens, increased aromatase expression will likely caused increased estrogen production

Table 1 Effects of EDCs on *cyp19a1b gene expression*, aromatase B protein expression and brain aromatase activity in different fish species

Compound	Species	Stage of exposure	Effects	References
Estradiol	*Danio rerio*	Larvae	↗ ARNm *cyp19a1b*	Menuet et al. (2005), Hinfray et al. (2006), Cheshenko et al. (2008)
		Larvae	↗ Aromatase B protein expression	Menuet et al. (2005), Cheshenko et al. (2008)
Ethinylestradiol	*Danio rerio*	Larvae	↗ ARNm *cyp19a1b*	Kazeto et al. (2004)
	Oryzias latipes	Adult	↗ Brain aromatase activity	Contractor et al. (2004)
	Leuciscus cephalus	Adult	↗ Brain aromatase activity	(Hinfray et al. 2010)
Para-nonylphenol	*Danio rerio*	Larvae	↗ ARNm *cyp19a1b*	Kazeto et al. (2004)
Genistein, α-zearalenol	*Danio rerio*	Larvae	↗ ARNm *cyp19a1b*	Le Page et al. (2010)
Testosterone, dihydrotestosterone	*Danio rerio*	Embryo	↗ *cyp19a1b* gene and protein expression	Mouriec et al. (2009b)
Flutamide	*Danio rerio*	Adult	ì Brain aromatase activity	Andersen et al. (2003)
Fadrozole	*Pimephales promela*	Adult	↘ Brain aromatase activity	Ankley et al. (2002)
	Oryzias latipes	Adult	↘ *cyp19a1b* expression	Zhang et al. (2008)
Androstatrienedione	*Danio rerio*	Adult	↘ Brain aromatase activity	Hinfray et al. (2006)
	Carassius auratus	Adult	↘ *cyp19a1b* expression	Gelinas et al. (1998)

within the RGCs and that these estrogens have the potential to affect the proliferative activity of RGCs and/or the fate of their progeny (Mouriec et al. 2008). Recent data support this assumption by showing that manipulation of estrogen levels or estrogen signalling indeed influences the cell proliferation in many brain regions, notably the pallium, subpallium, preoptic area and mediobasal hypothalamus (Diotel et al. unpublished). New evidences also support the view that the brain expresses the whole series of enzymes to *de novo* synthesize estrogens form cholesterol (Diotel et al. in preparation). The role of such neurosteroids is largely unknown, but it was suggested that they could play a role in brain sexualization. Indeed, in trout, steroidogenic enzymes exhibit a clear dimorphic pattern of expression at the time of gonadal differentiation (Vizziano-Cantonnet et al. 2011). Altogether, these data suggest that estrogens, in addition of

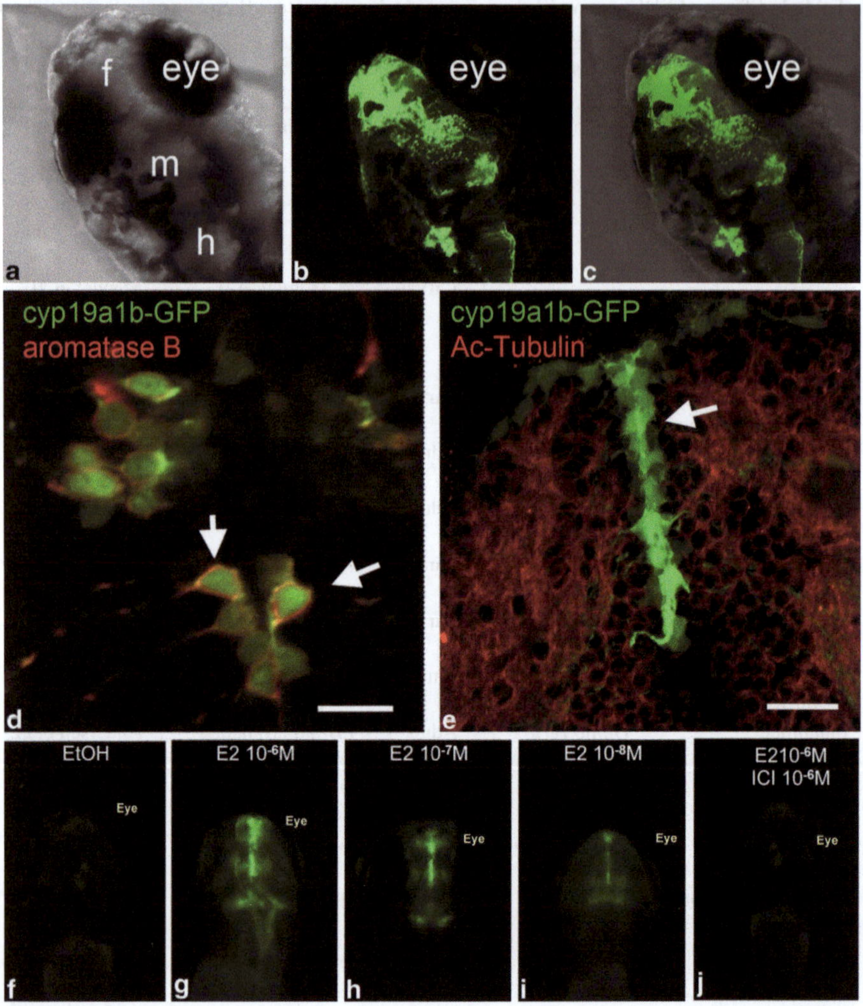

Fig. 2 (**a–c**) In vivo imaging of GFP expression in the brain of a transgenic (*tg*) (*cypa19a1b*-GFP). A 10 days old larva was treated with 10 nM estradiol, causing strong GFP expression in the radial glial cells of the forebrain (*f*), midbrain (*m*) and hindbrain (*h*). (**d**) Transgenic (*tg*) (*cypa19a1b*-GFP) zebrafish exhibit faithful GFP expression as shown by co-staining (*arroHs*) of GFP with aromatase B in radial glial cells only (confocal microscopy). Bar = 10 μm. (**e**) GFP expression is restricted to RGC (*green*) and is never in neurons stained in *red* with acetylated tubulin (Ac-tubulin) (confocal microscopy). Bar = 20 μm. (**f**) Exposure of zebrafish embryos for 2 days to increasing doses of E2 (**g–i**) causes a dose-dependent expression of GFP in the head. Controls treated with ethanol (EtOH) show no expression (**f**). The pure anti-estrogen ICI 182,780 blocks the effects of estradiol, showing the estrogen-receptor dependency of the E2 effects (**j**)

modulating classical neuroendocrine circuits, may participate in the construction of sexual dimorphic structures by influencing neurogenesis in a sexually-dimorphic manner.In mammals, aromatase is known for playing important functions on brain

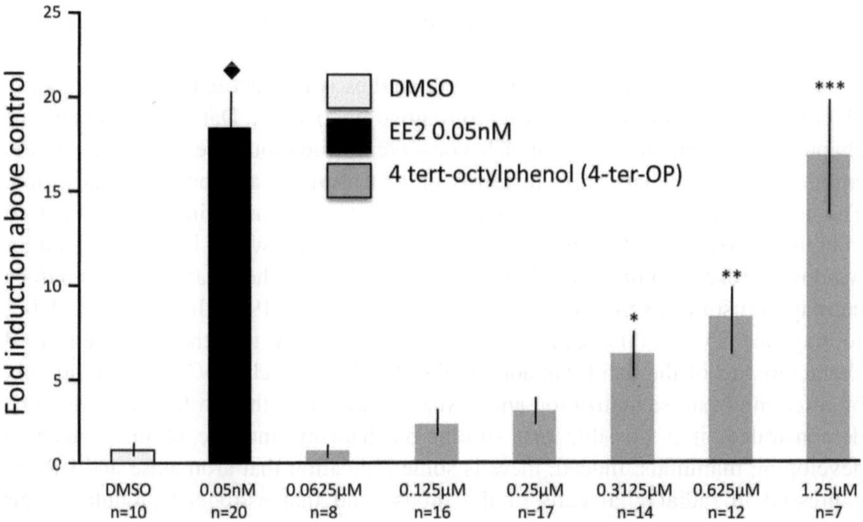

Fig. 3 Example of the use of tg(*cyp19a1b*-GFP) transgenic zebrafish to detect potential xenoestrogenic activity in vivo: Dose-dependent effect of 4-tert-Octylphenol exposure (5 days) on GFP-expression in the brain of zebrafish larvae. Zebrafish embryos were exposed for 5 days to a solution of 4-tert-octylphenol (0.0625–1.35 μM) dissolved in DMSO. Pictures of the brain were then taken using a fluorescence microscope and the intensity of the fluorescence was assessed by image analysis. Ethinylestradiol (EE2 0.05 nM) was used as a positive control. (Mean ± SEM; Significantly different from DMSO; $*p < 0.005$; $**p = 0.005$; $***p < 0.001$; ◆ $p < 0.0001$)

sexualization, however there are clear differences between fish and mammals. Teleost fish are indeed unique among vertebrate in having the capacity of changing sex in adults probably due to three characteristics of the adult brain: (i) an active neurogenesis, (ii) the persistence of radial glial progenitors in the whole brain and (iii) a very high aromatase activity associated to RGCs.

Although, we know very little about the mechanisms underlying sex change in the brain of fish, several studies point to a role for aromatase in this process (Tomy et al. 2007, 2009; Vizziano-Cantonnet et al. 2011). Furthermore, in the gilthead sea bream, a protogenous hermaphrodite species, the neurogenic activity of the caudal hypothalamus varies according to the sexual stage (Zikopoulos et al. 2001).

Conclusions and Perspectives

As could be expected, accumulating evidence indicates that a wide range of EDCs including estrogen mimics targets the brain of fish. Xenoestrogens, at environmentally relevant doses, affect GnRH neuron development with unknown consequences (Vosges et al. 2010). This is interesting in relation to observations in human populations showing an increase rate of sexual precocity in girls, an effect that could be linked to increased exposure to environmental estrogens (Rasier et al. 2006; Buck Louis et al. 2008).

Another main conclusion of the recent studies is the fact that fish significantly diverge from mammals with respect to brain aromatization. Data mentioned in this chapter show that the brain of fish conserves throughout life properties of the embryonic mammalian brain in terms of neurogenesis and brain sexualization. This is important as this suggests that, in contrast to mammals in which the period of high sensitivity to EDCs is the early development, fish would have a much larger window of susceptibility, including in adults. It is established that EDCs can induce hermaphrodism or even sex change in fish (Tyler et al. 1998; Jobling et al. 2002; Jobling and Tyler 2003; Coe et al. 2008). It is likely that the aforementioned characteristics of the fish brain add another level on which EDCs can act, notably by affecting brain sexualization and sexual behaviour.Although this remains to be demonstrated, it is possible that similar mechanisms may be of importance in developing mammals. Indeed, there is some indication that aromatase and ER are expressed in radial glial cells of the mouse and that estrogens modulate their proliferative activity (Martinez-Cerdeno et al. 2006). Other developmental effects of estrogens on the brain of rodents are suggested on the basis of studies on ARKO (Bakker 2003; Matsumoto et al. 2003; Pierman et al. 2008; Hill and Boon 2009; Bakker and Brock 2010) or ERKO (Wang et al. 2001, 2002) mice that exhibit a number of morphological and cognitive disorders. Clearly, this will deserve more investigation in the future.

Acknowledgements Unpublished data mentioned in this chapter were obtained with the support of the ANR NEED and the Post-Grenelle programme NEMO. The assistance of Dr. Denis Rouède with the confocal microscopy is greatly appreciated. We also thank the staff of the zebrafish transgenesis platform (INRA SCRIBE, IFR 140) for the help with the maintenance of fish.

References

Adolf B, Chapouton P, Lam CS, Topp S, Tannhauser B, Strahle U, Gotz M, Bally-Cuif L (2006) Conserved and acquired features of adult neurogenesis in the zebrafish telencephalon. Dev Biol 295:278–293

Allner B, von der Gonna S, Griebeler EM, Nikutowski N, Weltin A, Stahlschmidt-Allner P (2010) Reproductive functions of wild fish as bioindicators of reproductive toxicants in the aquatic environment. Environ Sci Pollut Res Int 17:505–518

Andersen L, Holbech H, Gessbo A, Norrgren L, Petersen GI (2003) Effects of exposure to 17alpha-ethinylestradiol during early development on sexual differentiation and induction of vitellogenin in zebrafish (*Danio rerio*). Comp Biochem Physiol C Toxicol Pharmacol 134:365–374

Ankley GT, Kahl MD, Jensen KM, Hornung MW, Korte JJ, Makynen EA, Leino RL (2002) Evaluation of the aromatase inhibitor fadrozole in a short-term reproduction assay with the fathead minnow (*Pimephales promelas*). Toxicol Sci 67:121–130

Bakker J (2003) Sexual differentiation of the neuroendocrine mechanisms regulating mate recognition in mammals. J Neuroendocrinol 15:615–621

Bakker J, Brock O (2010) Early oestrogens in shaping reproductive networks: evidence for a potential organisational role of oestradiol in female brain development. J Neuroendocrinol 22:728–735

Bentivoglio M, Mazzarello P (1999) The history of radial glia. Brain Res Bull 49:305–315

Boon WC, Chow JD, Simpson ER (2010) The multiple roles of estrogens and the enzyme aromatase. Prog Brain Res 181:209–232

Borg B, Andersson E, Mayer I, Lambert JG (1989) Aromatase activity in the brain of the three-spined stickleback, gasterosteus aculeatus. III. Effects of castration under different conditions and of replacement with different androgens. Exp Biol 48:149–152

Brandstatter R, Kotrschal K (1989) Life history of roach, *Rutilus rutilus* (Cyprinidae, Teleostei). A qualitative and quantitative study on the development of sensory brain areas. Brain Behav Evol 34:35–42

Brandstatter R, Kotrschal K (1990) Brain growth patterns in four European cyprinid fish species (Cyprinidae, Teleostei): roach (*Rutilus rutilus*), bream (*Abramis brama*), common carp (*Cyprinus carpio*) and sabre carp (*Pelecus cultratus*). Brain Behav Evol 35:195–211

Buck Louis GM, Gray LE Jr, Marcus M, Ojeda SR, Pescovitz OH, Witchel SF, Sippell W, Abbott DH, Soto A, Tyl RW, Bourguignon JP, Skakkebaek NE, Swan SH, Golub MS, Wabitsch M, Toppari J, Euling SY (2008) Environmental factors and puberty timing: expert panel research needs. Pediatrics 121(Suppl 3):S192–207

Callard GV, Pasmanik M (1987) The role of estrogen as a parahormone in brain and pituitary. Steroids 50:475–493

Callard GV, Petro Z, Ryan KJ (1978) Phylogenetic distribution of aromatase and other androgen-converting enzymes in the central nervous system. Endocrinology 103:2283–2290

Callard G, Schlinger B, Pasmanik M, Corina K (1990) Aromatization and estrogen action in brain. Prog Clin Biol Res 342:105–111

Callard GV, Tchoudakova AV, Kishida M, Wood E (2001) Differential tissue distribution, developmental programming, estrogen regulation and promoter characteristics of cyp19 genes in teleost fish. J Steroid Biochem Mol Biol 79:305–314

Cameron RS, Rakic P (1991) Glial cell lineage in the cerebral cortex: a review and synthesis. Glia 4:124–137

Campbell K, Gotz M (2002) Radial glia: Multi-purpose cells for vertebrate brain development. Trends Neurosci 25:235–238

Cheshenko K, Pakdel F, Segner H, Kah O, Eggen RI (2008) Interference of endocrine disrupting chemicals with aromatase CYP19 expression or activity, and consequences for reproduction of teleost fish. Gen Comp Endocrinol 155:31–62

Coe TS, Hamilton PB, Hodgson D, Paull GC, Stevens JR, Sumner K, Tyler CR (2008) An environmental estrogen alters reproductive hierarchies, disrupting sexual selection in group-spawning fish. Environ Sci Technol 42:5020–5025

Contractor RG, Foran CM, Li S, Willett KL (2004) Evidence of gender-and tissue-specific promoter methylation and the potential for ethinylestradiol-induced changes in Japanese medaka (*Oryzias latipes*) estrogen receptor and aromatase genes. J Toxicol Environ Health A 67:1–22

Costa MR, Gotz M, Berninger B (2010) What determines neurogenic competence in glia? Brain Res Rev 63:47–59

Diotel N, Vaillant C, Gueguen MM, Mironov S, Anglade I, Servili A, Pellegrini E, Kah O (2010a) Cxcr4 and Cxcl12 expression in radial glial cells of the brain of adult zebrafish. J Comp Neurol 518:4855–4876

Diotel N, Le Page Y, Mouriec K, Tong SK, Pellegrini E, Vaillant C, Anglade I, Brion F, Pakdel F, Chung BC, Kah O (2010) Aromatase in the brain of teleost fish: expression, regulation and putative functions. Front Neuroendocrinol 31(2):172–192

Do Rego JL, Seong JY, Burel D, Leprince J, Luu-The V, Tsutsui K, Tonon MC, Pelletier G, Vaudry H (2009) Neurosteroid biosynthesis: enzymatic pathways and neuroendocrine regulation by neurotransmitters and neuropeptides. Front Neuroendocrinol 30:259–301

Ekstrom P, Johnsson CM, Ohlin LM (2001) Ventricular proliferation zones in the brain of an adult teleost fish and their relation to neuromeres and migration (secondary matrix) zones. J Comp Neurol 436:92–110

Flouriot G, Pakdel F, Ducouret B, Valotaire Y (1995) Influence of xenobiotics on rainbow trout liver estrogen receptor and vitellogenin gene expression. J Mol Endocrinol 15:143–151

Forlano PM, Deitcher DL, Myers DA, Bass AH (2001) Anatomical distribution and cellular basis for high levels of aromatase activity in the brain of teleost fish: aromatase enzyme and mRNA expression identify glia as source. J Neurosci 21:8943–8955

Frye CA (2009) Steroids, reproductive endocrine function, and cognition. A review. Min Ginecol 61:563–585

Gelinas D, Pitoc GA, Callard GV (1998) Isolation of a goldfish brain cytochrome P450 aromatase cDNA: mRNA expression during the seasonal cycle and after steroid treatment. Mol Cell Endocrinol 138:81–93

Gonzalez A, Piferrer F (2003) Aromatase activity in the European sea bass (Dicentrarchus labrax L.) brain. Distribution and changes in relation to age, sex, and the annual reproductive cycle. Gen Comp Endocrinol 132:223–230

Grandel H, Kaslin J, Ganz J, Wenzel I, Brand M (2006) Neural stem cells and neurogenesis in the adult zebrafish brain: Origin, proliferation dynamics, migration and cell fate. Dev Biol 295:263–277

Hill RA, Boon WC (2009) Estrogens, brain, and behavior: lessons from knockout mouse models. Semin Reprod Med 27:218–228

Hinfray N, Palluel O, Turies C, Cousin C, Porcher JM, Brion F (2006) Brain and gonadal aromatase as potential targets of endocrine disrupting chemicals in a model species, the zebrafish (*Danio rerio*). Environ Toxicol 21:332–337

Hinfray N, Palluel O, Piccini B, Sanchez W, Ait-Aissa S, Noury P, Gomez E, Geraudie P, Minier C, Brion F, Porcher JM (2010) Endocrine disruption in wild populations of chub (*Leuciscus cephalus*) in contaminated French streams. Sci Total Environ 408:2146–2154

Hogan NS, Lean DR, Trudeau VL (2006) Exposures to estradiol, ethinylestradiol and octylphenol affect survival and growth of *Rana pipiens* and *Rana sylvatica* tadpoles. J Toxicol Environ Health A 69:1555–1569

Jagasia R, Song H, Gage FH, Lie DC (2006) New regulators in adult neurogenesis and their potential role for repair. Trends Mol Med 12:400–405

Jespersen A, Rasmussen TH, Hirche M, Sorensen KJ, Korsgaard B (2010) Effects of exposure to the xenoestrogen octylphenol and subsequent transfer to clean water on liver and gonad ultrastructure during early development of Zoarces viviparus embryos. J Exp Zool A Ecol Genet Physiol 313:399–409

Jobling S, Tyler CR (2003) Endocrine disruption, parasites and pollutants in wild freshwater fish. Parasitology 126(Suppl):S103–108

Jobling S, Coey S, Whitmore JG, Kime DE, Van Look KJ, McAllister BG, Beresford N, Henshaw AC, Brighty G, Tyler CR, Sumpter JP (2002) Wild intersex roach (*Rutilus rutilus*) have reduced fertility. Biol Reprod 67:515–524

Jobling S, Williams R, Johnson A, Taylor A, Gross-Sorokin M, Nolan M, Tyler CR, van Aerle R, Santos E, Brighty G (2006) Predicted exposures to steroid estrogens in U.K. rivers correlate with widespread sexual disruption in wild fish populations. Environ Health Perspect 114(Suppl 1):32–39

Kah O, Dufour S (2010) Conserved and divergent features of reproductive neuroendocrinology in fish. In: Norris DO, Lopez KH, Fishes (eds) Hormones and reproduction of vertebrates, vol 1. Elsevier, San Diego, pp 15–42

Kazeto Y, Place AR, Trant JM (2004) Effects of endocrine disrupting chemicals on the expression of CYP19 genes in zebrafish (*Danio rerio*) juveniles. Aquat Toxicol 69:25–34

Khan IA, Thomas P (2001) Disruption of neuroendocrine control of luteinizing hormone secretion by aroclor 1254 involves inhibition of hypothalamic tryptophan hydroxylase activity. Biol Reprod 64:955–964

Khan IA, Thomas P (2004) Aroclor 1254 inhibits tryptophan hydroxylase activity in rat brain. Arch Toxicol 78:316–320

Kidd KA, Blanchfield PJ, Mills KH, Palace VP, Evans RE, Lazorchak JM, Flick RW (2007) Collapse of a fish population after exposure to a synthetic estrogen. Proc Natl Acad Sci USA 104:8897–8901

Kirsche W (1967) On postembryonic matrix zones in the brain of various vertebrates and their relationship to the study of the brain structure. Z Mikrosk Anat Forsch 77:313–406

Konkle AT, McCarthy MM (2011) Developmental time course of estradiol, testosterone, and dihydrotestosterone levels in discrete regions of male and female rat brain. Endocrinology 152:223–235

Kranz D, Richter W (1970a) Autoradiographic studies on the synthesis of DNA in the cerebellum and medulla oblongata of teleosts of various ages. Z Mikrosk Anat Forsch 82:264–292

Kranz D, Richter W (1970b) Autoradiographic studies on the localization of the matrix zones of the diencephalon of young and adult *Lebistes reticulatus* (Teleostae). Z Mikrosk Anat Forsch 82:42–66

Kriegstein A, Alvarez-Buylla A (2009) The glial nature of embryonic and adult neural stem cells. Annu Rev Neurosci 32:149–184

Le Page Y, Scholze M, Kah O, Pakdel F (2006) Assessment of xenoestrogens using three distinct estrogen receptors and the zebrafish brain aromatase gene in a highly responsive glial cell system. Environ Health Perspect 114:752–758

Le Page Y, Menuet A, Kah O, Pakdel F (2008) Characterization of a cis-acting element involved in cell-specific expression of the zebrafish brain aromatase gene. Mol Reprod Dev 75:1549–1557

Le Page Y, Diotel N, Vaillant C, Pellegrini E, Anglade I, Mérot Y, Kah O (2010) Aromatase, brain sexualization and plasticity: the fish paradigm. Eur J Neurosci 32:2105–2115

Lindsey BW, Tropepe V (2006) A comparative framework for understanding the biological principles of adult neurogenesis. Prog Neurobiol 80:281–307

Marsh KE, Creutz LM, Hawkins MB, Godwin J (2006) Aromatase immunoreactivity in the bluehead wrasse brain, *Thalassoma bifasciatum*: immunolocalization and co-regionalization with arginine vasotocin and tyrosine hydroxylase. Brain Res 1126:91–101

Martinez-Cerdeno V, Noctor SC, Kriegstein AR (2006) Estradiol stimulates progenitor cell division in the ventricular and subventricular zones of the embryonic neocortex. Eur J Neurosci 24:3475–3488

Martinez-Cerdeño V, Noctor SC, Kriegstein AR (2006) Estradiol stimulates progenitor cell division in the ventricular and subventricular zones of the embryonic neocortex. Eur J Neurosci 24:3475–3488

März M, Chapouton P, Diotel N, Vaillant C, Hesl B, Takamiya M, Lam CS, Kah O, Bally-Cuif L, Strahle U (2010) Heterogeneity in progenitor cell subtypes in the ventricular zone of the zebrafish adult telencephalon. Glia 58:870–888

Matsumoto T, Honda S, Harada N (2003) Neurological effects of aromatase deficiency in the mouse. J Steroid Biochem Mol Biol 86:357–365

McEwen BS (2001) Invited review: Estrogens effects on the brain: Multiple sites and molecular mechanisms. J Appl Physiol 91:2785–2801

Mennigen JA, Harris EA, Chang JP, Moon TW, Trudeau VL (2009) Fluoxetine affects weight gain and expression of feeding peptides in the female goldfish brain. Regul Pept 155:99–104

Menuet A, Anglade I, Le Guevel R, Pellegrini E, Pakdel F, Kah O (2003) Distribution of aromatase mRNA and protein in the brain and pituitary of female rainbow trout: comparison with estrogen receptor alpha. J Comp Neurol 462:180–193

Menuet A, Pellegrini E, Brion F, Gueguen MM, Anglade I, Pakdel F, Kah O (2005) Expression and estrogen-dependent regulation of the zebrafish brain aromatase gene. J Comp Neurol 485:304–320

Mouriec K, Pellegrini E, Anglade I, Menuet A, Adrio F, Thieulant ML, Pakdel F, Kah O (2008) Synthesis of estrogens in progenitor cells of adult fish brain: evolutive novelty or exaggeration of a more general mechanism implicating estrogens in neurogenesis? Brain Res Bull 75 (2-4):274–280

Mouriec K, Gueguen MM, Manuel C, Percevault F, Thieulant ML, Pakdel F, Kah O (2009a) Androgens upregulate cyp19a1b (aromatase B) gene expression in the brain of zebrafish (*Danio rerio*) through estrogen receptors. Biol Reprod 80:889–896

Mouriec K, Lareyre JJ, Tong SK, Le Page Y, Vaillant C, Pellegrini E, Pakdel F, Chung BC, Kah O, Anglade I (2009) Early regulation of brain aromatase (cyp19a1b) by estrogen receptors during zebrafish development. Dev Dyn 238(10):2641–2651

Nagahama Y, Nakamura M, Kitano T, Tokumoto T (2004) Sexual plasticity in fish: A possible target of endocrine disruptor action. Environ Sci 11:73–82

Ninkovic J, Gotz M (2007) Signaling in adult neurogenesis: from stem cell niche to neuronal networks. Curr Opin Neurobiol 17(3):338–44

Noctor SC, Flint AC, Weissman TA, Dammerman RS, Kriegstein AR (2001) Neurons derived from radial glial cells establish radial units in neocortex. Nature 409:714–720

Oakes KD et al (2010) Environmental risk assessment for the serotonin re-uptake inhibitor fluoxetine: Case study using the European risk assessment framework. Integr Environ Assess Manag 6(Suppl):524–539

Pasmanik M, Callard GV (1985) Aromatase and 5 alpha-reductase in the teleost brain, spinal cord, and pituitary gland. Gen Comp Endocrinol 60:244–251

Pellegrini E, Mouriec K, Anglade I, Menuet A, Le Page Y, Gueguen MM, Marmignon MH, Brion F, Pakdel F, Kah O (2007) Identification of aromatase-positive radial glial cells as progenitor cells in the ventricular layer of the forebrain in zebrafish. J Comp Neurol 501:150–167

Petit F, Le Goff P, Cravedi JP, Valotaire Y, Pakdel F (1997) Two complementary bioassays for screening the estrogenic potency of xenobiotics: Recombinant yeast for trout estrogen receptor and trout hepatocyte cultures. J Mol Endocrinol 19:321–335

Pierman S, Douhard Q, Bakker J (2008) Evidence for a role of early oestrogens in the central processing of sexually relevant olfactory cues in female mice. Eur J Neurosci 27:423–431

Rakic P (1978) Neuronal migration and contact guidance in the primate telencephalon. Postgrad Med J 54(Suppl 1):25–40

Rakic P (2003) Elusive radial glial cells: Historical and evolutionary perspective. Glia 43:19–32

Rasier G, Toppari J, Parent AS, Bourguignon JP (2006) Female sexual maturation and reproduction after prepubertal exposure to estrogens and endocrine disrupting chemicals: a review of rodent and human data. Mol Cell Endocrinol 254–255:187–201

Rowitch DH, Kriegstein AR (2010) Developmental genetics of vertebrate glial-cell specification. Nature 468:214–222

Santen RJ, Brodie H, Simpson ER, Siiteri PK, Brodie A (2009) History of aromatase: saga of an important biological mediator and therapeutic target. Endocr Rev 30:343–375

Sharpe RM, Skakkebaek NE (2008) Testicular dysgenesis syndrome: mechanistic insights and potential new downstream effects. Fertil Steril 89:e33–38

Simpson ER (2004) Aromatase: biologic relevance of tissue-specific expression. Semin Reprod Med 22:11–23

Strobl-Mazzulla PH, Moncaut NP, Lopez GC, Miranda LA, Canario AV, Somoza GM (2005) Brain aromatase from pejerrey fish (*Odontesthes bonariensis*): cDNA cloning, tissue expression, and immunohistochemical localization. Gen Comp Endocrinol 143:21–32

Strobl-Mazzulla PH, Lethimonier C, Gueguen MM, Karube M, Fernandino JI, Yoshizaki G, Patino R, Strussmann CA, Kah O, Somoza GM (2008) Brain aromatase (Cyp19A2) and estrogen receptors, in larvae and adult pejerrey fish *Odontesthes bonariensis*: Neuroanatomical and functional relations. Gen Comp Endocrinol 158:191–201

Sugiyama N, Barros RP, Warner M, Gustafsson JA (2010) ERbeta: recent understanding of estrogen signaling. Trends Endocrinol Metab 21(9):545–52

Sumpter JP, Jobling S (1995) Vitellogenesis as a biomarker for estrogenic contamination of the aquatic environment. Environ Health Perspect 103(Suppl 7):173–178

Taverna E, Huttner WB (2010) Neural progenitor nuclei IN motion. Neuron 67:906–914

Tchoudakova A, Callard GV (1998) Identification of multiple CYP19 genes encoding different cytochrome P450 aromatase isozymes in brain and ovary. Endocrinology 139:2179–2189

Thomas P, Rahman MS, Khan IA, Kummer JA (2007) Widespread endocrine disruption and reproductive impairment in an estuarine fish population exposed to seasonal hypoxia. Proc Biol Sci 274:2693–2701

Tomy S, Wu GC, Huang HR, Dufour S, Chang CF (2007) Developmental expression of key steroidogenic enzymes in the brain of protandrous black porgy fish, *Acanthopagrus schlegeli*. J Neuroendocrinol 19:643–655

Tomy S, Wu GC, Huang HR, Chang CF (2009) Age-dependent differential expression of genes involved in steroid signalling pathway in the brain of protandrous black porgy, *Acanthopagrus schlegeli*. Dev Neurobiol 69:299–313

Tong SK, Mouriec K, Kuo MW, Pellegrini E, Guguen MM, Brion F, Kah O, Chung BC (2009) A Cyp19a1b-gfp (aromatase B) transgenic zebrafish line that expresses GFP in radial glial cells. Genesis 47(2):67–73

Toran-Allerand CD (2004a) Estrogen and the brain: beyond ER-alpha and ER-beta. Exp Gerontol 39:1579–1586

Toran-Allerand CD (2004b) Minireview: a plethora of estrogen receptors in the brain: where will it end? Endocrinology 145:1069–1074

Tyler CR, Jobling S, Sumpter JP (1998) Endocrine disruption in wildlife: a critical review of the evidence. Crit Rev Toxicol 28:319–361

Vaillant C, Le Guellec C, Pakdel F, Valotaire Y (1988) Vitellogenin gene expression in primary culture of male rainbow trout hepatocytes. Gen Comp Endocrinol 70:284–290

Vizziano-Cantonnet D, Anglade I, Pellegrini E, Gueguen MM, Fostier A, Guiguen Y, Kah O (2011) Sexual dimorphism in the brain aromatase expression and activity, and in the central expression of other steroidogenic enzymes during the period of sex differentiation in monosex rainbow trout populations. Gen Comp Endocrinol 170:346–355

Vosges M, Le Page Y, Chung BC, Combarnous Y, Porcher JM, Kah O, Brion F (2010) 17alpha-ethinylestradiol disrupts the ontogeny of the forebrain GnRH system and the expression of brain aromatase during early development of zebrafish. Aquat Toxicol 99:479–491

Walf AA, Frye CA (2006) A review and update of mechanisms of estrogen in the hippocampus and amygdala for anxiety and depression behavior. Neuropsychopharmacology 31:1097–1111

Wang L, Andersson S, Warner M, Gustafsson JA (2001) Morphological abnormalities in the brains of estrogen receptor beta knockout mice. Proc Natl Acad Sci USA 98:2792–2796

Wang L, Andersson S, Warner M, Gustafsson JA (2002) Estrogen actions in the brain. Sci STKE 2002:pe29

Zhang X, Hecker M, Park JW, Tompsett AR, Jones PD, Newsted J, Au DW, Kong R, Wu RS, Giesy JP (2008) Time-dependent transcriptional profiles of genes of the hypothalamic-pituitary-gonadal axis in medaka (*Oryzias latipes*) exposed to fadrozole and 17beta-trenbolone. Environ Toxicol Chem 27:2504–2511

Zikopoulos B, Kentouri M, Dermon CR (2001) Cell genesis in the hypothalamus is associated to the sexual phase of a hermaphrodite teleost. Neuroreport 12:2477–2481

Zupanc GK (2001) Adult neurogensis and neuronal regeneration in the central nervous system of teleost fish. Brain Behav Evol 58(5):250–275

Zupanc GK, Zupanc MM (2006) New neurons for the injured brain: mechanisms of neuronal regeneration in adult teleost fish. Regen Med 1(2):207–216

Zupanc GK, Hinsch K, Gage FH (2005) Proliferation, migration, neuronal differentiation, and long-term survival of new cells in the adult zebrafish brain. J Comp Neurol 488:290–319

Exposure to Environmental Chemicals as a Risk Factor for Diabetes Development

P. Grandjean

Abstract The increasing prevalence of type 2 diabetes mellitus has been linked primarily to obesity and lack of exercise, but certain environmental chemicals can induce both insulin resistance and disruption of insulin secretion in experimental models. Epidemiologic studies support a role for environmental chemical exposure in the development of type 2 diabetes. However, epidemiological studies cannot settle the question of causative substances, as environmental exposures include complex mixtures of persistent halogenated compounds, which are mutually correlated. In vitro studies suggest differences in toxic profiles: for example, dioxin is toxic to beta cells, but has only weak effects on insulin resistance. A conundrum in the pathogenesis is the direction of changes in fasting insulin concentrations at increased exposures. An increase in fasting insulin would suggest a physiological response to insulin resistance, whereas a decrease could be a consequence of beta cell toxicity. In populations of exposed adults, dioxin exposure has shown a positive correlation with fasting insulin. As diabetes is particularly common in the elderly, who may have accumulated large body burdens of persistent environmental chemicals, we examined this issue in elderly Faroese with high pollutant exposures from traditional diets of pilot whale blubber. In 713 subjects aged 70–74 years (64% of eligible population), we estimated lifetime exposure levels from the past frequencies of dinners with pilot whale and other traditional food and by analysis of serum for polychlorinated biphenyls (PCBs) and related substances. Septuagenarians with type 2 diabetes or impaired fasting glycaemia tended to have higher PCB concentrations and higher past intakes of traditional food, especially during childhood and adolescence. In non-diabetic subjects, the fasting insulin concentration decreased by 7% (95% CI: −12, −2.1) for each doubling of the PCB concentration after adjustment for sex and body mass index. Conversely, an increase

P. Grandjean (✉)
Department of Environmental Medicine, University of Southern Denmark, Odense, Denmark

Department of Environmental Health, Harvard School of Public Health, Boston, MA, USA
e-mail: pgrandjean@health.sdu.dk

J.-P. Bourguignon et al. (eds.), *Multi-System Endocrine Disruption*,
Research and Perspectives in Endocrine Interactions,
DOI 10.1007/978-3-642-22775-2_6, © Springer-Verlag Berlin Heidelberg 2011

of the same magnitude was seen in the fasting glucose concentration. Impaired insulin secretion may therefore constitute an important part of the type 2 diabetes pathogenesis associated with exposure to persistent lipophilic food contaminants. Intensified abatement of exposure to relevant endocrine disruptors should be considered an attractive means of complementing preventive efforts against type 2 diabetes.

Introduction

The increasing prevalence of type 2 diabetes mellitus has been linked primarily to obesity and lack of exercise (World Health Organization (WHO) 2010). However, experimental evidence suggests that exposure to dioxins and other persistent halogenated chemicals may result in both insulin resistance and disruption of insulin secretion (Carpenter 2008; Ruzzin et al. 2010).

Epidemiologic studies also support a role for environmental chemicals in the pathogenesis of type 2 diabetes. Thus, in subjects with high-level exposure to dioxins and related substances, an increased risk of developing diabetes has been documented, in some cases more clearly in women (Consonni et al. 2008; Henriksen et al. 1997; Ukropec et al. 2010; Wang et al. 2008). Within the general population, diabetes patients have higher serum concentrations of persistent, lipophilic pollutants, including polychlorinated biphenyls (PCBs) and the pesticide metabolite 2,2-bis(4-chlorophenyl)-1,1-dichloroethene (DDE); Everett and Matheson 2010; Lee et al. 2006, 2007; Patel et al. 2010; Rignell-Hydbom et al. 2007, 2009).

In addition, subjects with impaired fasting glycemia tend to have higher cumulated pollutant concentrations in serum (Cranmer et al. 2000; Kern et al. 2004). Adult Greenland Inuit showed an inverse relation between the serum PCB concentration and the insulin concentration at 2h after glucose challenge (Jørgensen et al. 2008). Likewise, in US veterans with existing diabetes, fasting insulin concentrations decreased at higher exposures (Michalek et al. 1999). However, non-diabetic veterans at an average age of 53 years showed higher fasting serum-insulin concentrations at increased dioxin exposure levels (Longnecker and Michalek 2000). Although somewhat equivocal, this evidence suggests that exposures to persistent environmental chemicals may play an important role in the pathogenesis of the current diabetes epidemic. Given the huge public health implications of the massive increase in diabetes prevalence, the potential for disease prevention through targeted pollution abatement deserves attention.

Endocrine Disruption in Diabetes Pathogenesis

Experimental models suggest that certain endocrine disrupting chemicals present in environmental pollution can cause obesity, metabolic syndrome, and insulin resistance (Newbold et al. 2008). Several mechanisms may be involved in the

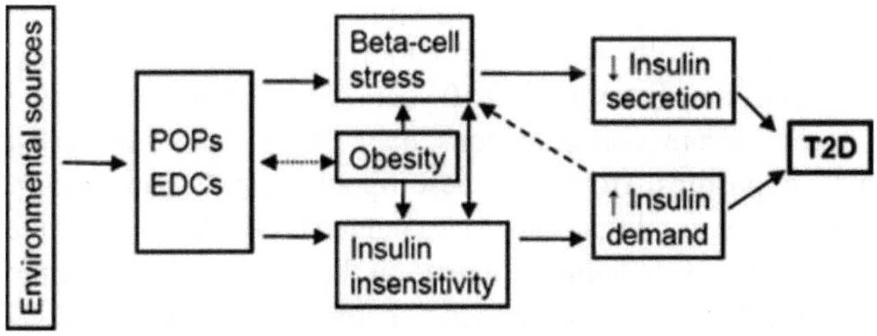

Fig. 1 Possible pathogenesis of type 2 diabetes (T2D) induced by environmental exposure to persistent organic pollutants (POPs) and endocrine disrupting chemicals (EDCs)

pathogenesis of type 2 diabetes (Fig. 1). While insulin resistance may be due to inflammation or lipid overload of fat cells, mechanisms of beta cell toxicity include activation of the intranuclear aryl hydrocarbon receptor (Kim et al. 2009; Kurita et al. 2009), interaction with the estrogen receptor (Nadal et al. 2009), or induction of peroxidation reactions (Martino et al. 2009). In the short term, beta cell toxicity results in increased insulin release, only later followed by a decrease as the cells wither. Increased insulin release has also been observed in whole animal models (Ruzzin et al. 2010), in this case perhaps more likely due to compensation for insulin resistance.

While beta cell toxicity due to organohalogen compounds has not yet been systematically addressed, induction of insulin resistance by these substances has been studied in a cell line system based on pre-adipocytes/adipocytes to identify cell-autonomous effects. Ruzzin et al. (2010) added the organohalogen compounds at physiologically relevant concentrations and controlled the concomitant insulin exposures. Using relative concentrations similar to salmon oil, the effects of dioxins were negligible compared to the much stronger effects of other substances, such as DDE and PCBs (Ruzzin et al. 2010).

The relevance of these observations is emphasized by the demonstration in rodents that contaminant levels present in salmon oil can induce the development of T2D. While both the crude and the refined fish oil cleaned of organohalogen contaminants caused obesity, it was only the crude oil with the contaminants that triggered the development of insulin resistance (Ruzzin et al. 2010). Clearly, these environmental chemicals must be strongly suspected of playing a role in the disease pathogenesis.

Epidemiological Linkage of Chemical Exposures to Diabetes Causation

The pathogenesis of type 2 diabetes may originate decades before the clinical diagnosis is made (Jones and Ozanne 2009). Thus, some general population studies have analyzed banked serum samples obtained up to several years before the appearance of clinical abnormalities (Longnecker and Michalek 2000; Rignell-Hydbom et al. 2009). Still, as the elimination half-time of dioxins, certain PCB congeners and DDE may be as long as 10 years (Grandjean et al. 2008; Seegal et al. 2010), concurrent serum concentrations may provide a reasonable estimate of cumulated past exposures. However, these lipophilic pollutants accumulate in fat tissue, and subjects with a high body mass index (BMI) therefore dilute their body burden within a larger distribution volume; on the other hand, with time, storage in body fat leads to a longer retention of the chemicals (Wolff et al. 2007). Still, the elimination half-life as such does not seem to be affected by the development of type 2 diabetes (Michalek et al. 2003). In addition, the known BMI-associated increase in diabetes risk has been found to be greater in subjects with an increased serum concentration of the pollutants (Lee et al. 2006), thus calling for assessment of possible interactions.

In searching for clues to shed new light on the possible etiologic role of lipophilic pollutants, we have examined abnormalities of glucose metabolism in Faroese elderly with a high risk of diabetes development (Grandjean et al. 2011). The Faroe Islands community offers a unique research opportunity, as pollutant exposures primarily originate from traditional diets that include seabirds and blubber of the pilot whale (Weihe et al. 2008). Past exposures may therefore be assessed by dietary questionnaire as well as by serum analysis. The 168 septuagenarians (24% of the cohort) who had type 2 diabetes and the 78 additional subjects (11%) with impaired fasting glycemia tended to show higher serum concentrations of DDE and PCB, especially in subjects with a BMI close to the average. The association was much weaker in obese subjects and those who were in the lowest BMI quartile. The BMI-dependent associations suggest that, in subjects who have already developed type 2 diabetes or impaired fasting glycemia, the serum concentration of persistent diabetogenic chemicals may no longer reflect the causative level of exposure.

We therefore focused on the septuagenarians with a normal fasting glucose concentration and normal glycated hemoglobin levels. In these subjects, the fasting plasma insulin concentration decreased substantially at higher PCB and DDE levels, both in all non-diabetic subjects ($N = 543$) and in subjects with normal glucose tolerance ($N = 466$). Again, the effect was about twice as strong in subjects with a high BMI. At the same time, the fasting glucose concentration increased at higher exposures, although in this group not exceeding a level of 6.9 mmol/L, which would have triggered a diabetes diagnosis. Further support comes from calculation of the homeostasis model assessment (HOMA) beta index from the ratio of the fasting plasma concentrations of insulin and glucose; this index also substantially decreased at higher exposure levels. In addition, questionnaire data on traditional dietary habits during childhood and young adulthood supported these associations, although the associations were less robust (Grandjean et al. 2011).

In interpreting results from this study and several others with similar or related findings, one must note the caveats associated with measurement of persistent environmental chemicals as biomarkers of long-term accumulation (Wolff et al.

2007). Differences in BMI may impact on serum concentrations and the independent effect of obesity on diabetes risk may further cloud any diabetogenic effect reflected by the current serum PCB concentration.

Decreased fasting plasma-insulin in highly exposed subjects has been reported previously (Jørgensen et al. 2008; Michalek et al. 1999), although a positive association with increased dioxin exposures has been observed in younger adults (Cranmer et al. 2000; Longnecker and Michalek 2000). These findings may appear to be contradictory, but they may also reflect different stages of the pathogenesis of type 2 diabetes mellitus. In the initial phase, early insulin resistance may be compensated for by increases in insulin secretion with the aim of maintaining normal plasma glucose concentrations. With time, exhaustion of the beta cells develops as an indirect result of the insulin resistance (Beck-Nielsen et al. 2003). The decline of beta cell function may be accelerated by toxicity due to environmental chemicals. Accordingly, the dioxin-associated increases in fasting plasma insulin concentrations in younger subjects may reflect initial stimulation of insulin secretion, as seen in experimental models, perhaps triggered by early insulin resistance. Later on, the trend in fasting insulin concentration may be reversed as a sign of beta cell depletion and toxic stresses. This change constitutes a further step in the development of type 2 diabetes mellitus and may be serious if coinciding with obesity. Thus, the overall findings suggest that exposure to organohalogen substances may lead to both insulin resistance and to beta cell toxicity, as also indicated by experimental studies (Ruzzin et al. 2010).

Uncertainties in Current Knowledge

The evidence summarized so far relies on a variety of laboratory models and observational studies in humans. In general, the validity of the experimental designs is incompletely known, and all of the epidemiological data originate from cross-sectional studies, not from randomized trials or even from long-term prospective studies of exposed populations. Thus, the information available is far from complete and allows only tentative conclusions.

An important weakness is that most epidemiological studies rely on a single measurement of serum concentrations in a cross-sectional design or with a follow-up that rarely spans more than a decade. The serum concentration is likely to be an imprecise indicator of the life-time exposure to the causative contaminants, and it may not reflect the exposure level at the most vulnerable time window for diabetes development, in part because differences and changes in BMI may impact on the serum concentrations of the persistent pollutants. Such imprecision would tend to cause a bias towards the null, thereby underestimating the effect. Similar concerns relate to the use of questionnaire answers on past diets, and they are further exacerbated by the absence of information on concentrations of contaminants in traditional food in the past. While some populations had a peak exposure during adulthood in connection with military or occupational activities, general population groups have experienced lower and long-term exposures. The discovery of PCBs and other persistent organic pollutants in the environment dates back to the 1970s, but substantial amounts probably began to enter the environment as early as the 1950s. Few subjects included in the studies reported so far would have been exposed to any substantial degree prenatally. The present evidence therefore does not address the effects of diabetogenic substances caused by developmental exposures.

Assuming that the association between organohalogen exposure and development of type 2 diabetes is causal, the role of individual substances or groups of, e.g., PCB congeners, in the disease etiology needs to be assessed. Unfortunately, most populations are exposed to mixtures of potentially diabetogenic substances, often with a high correlation between their serum concentrations. For example, in a US population-based study, serum concentrations of six different pollutants detectable in at least 80% of the participants showed, based on their sum, an adjusted odds ratio of 38 for diabetes in subjects in the highest exposure quintile, as compared with those in the lowest (Lee et al. 2006). All pollutants seemed to contribute to the increased risk. However, all of them are also highly persistent, and they may merely serve as markers of past as well as cumulated exposures to environmental chemicals in a more general sense, and some of the causative substance may conceivably have been eliminated over time. The degree of statistical significance of such correlations cannot be interpreted as support for causation.

In addition to dioxin-related substances, PCB and DDE, other diabetogenic candidates among the persistent environmental chemicals include the brominated flame retardants (Lim et al. 2008) and perfluorinated compounds (Lin et al. 2009). However, exposures to many of these persistent pollutants are interrelated to such an extent that their relative contributions cannot be ascertained from current evidence. Accordingly, the identity and specific effects of the potentially causative substance(s) remain unclear. Nonetheless, one conclusion seems to emerge in regard to the dose–response relationships. The odds ratio of 38 obtained in the US population exposed to background concentrations of organohalogen substances (Lee et al. 2006) is much greater than the odds ratios obtained in other populations at higher exposures (Grandjean et al. 2011; Turyk et al. 2009; Ukropec et al. 2010; Wang et al. 2008). The apparent decrease in diabetes risks at higher exposures indicates that the diabetogenicity may not increase linearly with the dose. Still, as the odds ratios may have been affected by small numbers of subjects, this issue deserves to be explored in further detail.

Public Health Perspective

The increasing world-wide prevalence of type 2 diabetes is in part a result of an increased survival of the diabetic population due to improved treatment, but it is also a result of an unhealthy, sedentary lifestyle and an energy-dense diet, both resulting in increased body weight (World Health Organization (WHO) 2010).

If environmental pollutants play a role in triggering type 2 diabetes, prevention may need to involve mechanisms other than individual intervention, as these substances are generally invisible to the consumer and may be difficult to address within current health care systems. The contaminants studied so far have now largely been banned, and exposures have since decreased. However, large cohorts were born around the middle of the twentieth century and are now approaching the age with the highest diabetes incidence rates. These subjects were born at a time when environmental accumulation of potential diabetogenic pollutants was at a peak, thus possibly causing interference with the developmental programming of glucose metabolism (Jones and Ozanne 2009). Further research is called for to elucidate the role of persistent organohalogen pollutants in human diabetes pathogenesis. A combination of systematic experimental studies along with large-scale prospective examinations of exposed populations would be desirable. Intensified prevention of exposure to environmental pollutants should be considered as an attractive means of complementing preventive efforts against type 2 diabetes. Such efforts should not be unduly delayed because of insufficient evidence. While much still remains to be understood, current evidence most likely underestimates the true extent of the public health consequences of environmental dissemination of diabetogenic chemicals.

References

Beck-Nielsen H, Vaag A, Poulsen P, Gaster M (2003) Metabolic and genetic influence on glucose metabolism in type 2 diabetic subjects–experiences from relatives and twin studies. Best Pract Res Clin Endocrinol Metab 17:445–467

Carpenter DO (2008) Environmental contaminants as risk factors for developing diabetes. Rev Environ Health 2008(23):59–74

Consonni D, Pesatori AC, Zocchetti C, Sindaco R, D'Oro LC, Rubagotti M, Bertazzi PA (2008) Mortality in a population exposed to dioxin after the Seveso, Italy, accident in 1976: 25 years of follow-up. Am J Epidemiol 167:847–858

Cranmer M, Louie S, Kennedy RH, Kern PA, Fonseca VA (2000) Exposure to 2,3,7,8-tetrachlorodibenzo-p-dioxin (TCDD) is associated with hyperinsulinemia and insulin resistance. Toxicol Sci 56:431–436

Everett CJ, Matheson EM (2010) Biomarkers of pesticide exposure and diabetes in the 1999–2004 national health and nutrition examination survey. Environ Int 36:398–401

Grandjean P, Budtz-Jorgensen E, Barr DB, Needham LL, Weihe P, Heinzow B (2008) Elimination half-lives of polychlorinated biphenyl congeners in children. Environ Sci Technol 42:6991–6996

Grandjean P, Henriksen JE, Choi AL, Petersen MS, Dalgard C, Nielsen F, Weihe P (2011) Exposure to marine food pollutants as a risk factor for hypoinsulinemia and type 2 diabetes. Epidemiology 22:410–417

Henriksen GL, Ketchum NS, Michalek JE, Swaby JA (1997) Serum dioxin and diabetes mellitus in veterans of Operation Ranch Hand. Epidemiology 8:252–258

Jones RH, Ozanne SE (2009) Fetal programming of glucose-insulin metabolism. Mol Cell Endocrinol 297:4–9

Jørgensen ME, Borch-Johnsen K, Bjerregaard P (2008) A cross-sectional study of the association between persistent organic pollutants and glucose intolerance among Greenland Inuit. Diabetologia 51:1416–1422

Kern PA, Said S, Jackson WG Jr, Michalek JE (2004) Insulin sensitivity following agent orange exposure in Vietnam veterans with high blood levels of 2,3,7,8-tetrachlorodibenzo-p-dioxin. J Clin Endocrinol Metab 89:4665–4672

Kim YH, Shim YJ, Shin YJ, Sul D, Lee E, Min BH (2009) 2,3,7,8-Tetrachlorodibenzo-p-dioxin (TCDD) induces calcium influx through T-type calcium channel and enhances lysosomal exocytosis and insulin secretion in INS-1 cells. Int J Toxicol 28:151–161

Kurita H, Yoshioka W, Nishimura N, Kubota N, Kadowaki T, Tohyama C (2009) Aryl hydrocarbon receptor-mediated effects of 2,3,7,8-tetrachlorodibenzo-p-dioxin on glucose-stimulated insulin secretion in mice. J Appl Toxicol 29:689–694

Lee DH, Lee IK, Song K, Steffes M, Toscano W, Baker BA, Jacobs DR Jr (2006) A strong dose–response relation between serum concentrations of persistent organic pollutants and diabetes: results from the National Health and Examination Survey 1999–2002. Diabetes Care 29:1638–1644

Lee DH, Lee IK, Porta M, Steffes M, Jacobs DR Jr (2007) Relationship between serum concentrations of persistent organic pollutants and the prevalence of metabolic syndrome among non-diabetic adults: results from the National Health and Nutrition Examination Survey 1999–2002. Diabetologia 50:1841–1851

Lim JS, Lee DH, Jacobs DR Jr (2008) Association of brominated flame retardants with diabetes and metabolic syndrome in the U.S. population, 2003–2004. Diabetes Care 31:1802–1807

Lin CY, Chen PC, Lin YC, Lin LY (2009) Association among serum perfluoroalkyl chemicals, glucose homeostasis, and metabolic syndrome in adolescents and adults. Diabetes Care 32:702–707

Longnecker MP, Michalek JE (2000) Serum dioxin level in relation to diabetes mellitus among Air Force veterans with background levels of exposure. Epidemiology 11:44–48

Martino L, Novelli M, Masini M, Chimenti D, Piaggi S, Masiello P, De Tata V (2009) Dehydroascorbate protection against dioxin-induced toxicity in the beta-cell line INS-1E. Toxicol Lett 189:27–34

Michalek JE, Akhtar FZ, Kiel JL (1999) Serum dioxin, insulin, fasting glucose, and sex hormone-binding globulin in veterans of Operation Ranch Hand. J Clin Endocrinol Metab 84:1540–1543

Michalek JE, Ketchum NS, Tripathi RC (2003) Diabetes mellitus and 2,3,7,8-tetrachlorodibenzo-p-dioxin elimination in veterans of Operation Ranch Hand. J Toxicol Environ Health A 66:211–221

Nadal A, Alonso-Magdalena P, Soriano S, Quesada I, Ropero AB (2009) The pancreatic beta-cell as a target of estrogens and xenoestrogens: implications for blood glucose homeostasis and diabetes. Mol Cell Endocrinol 304:63–68

Newbold RR, Padilla-Banks E, Jefferson WN, Heindel JJ (2008) Effects of endocrine disruptors on obesity. Int J Androl 31:201–208

Patel CJ, Bhattacharya J, Butte AJ (2010) An environment-wide association study (EWAS) on type 2 diabetes mellitus. PLoS One 5:e10746

Rignell-Hydbom A, Rylander L, Hagmar L (2007) Exposure to persistent organochlorine pollutants and type 2 diabetes mellitus. Hum Exp Toxicol 26:447–452

Rignell-Hydbom A, Lidfeldt J, Kiviranta H, Rantakokko P, Samsioe G, Agardh CD, Rylander L (2009) Exposure to p, p′-DDE: a risk factor for type 2 diabetes. PLoS One 4:e7503

Ruzzin J, Petersen R, Meugnier E, Madsen L, Lock EJ, Lillefosse H, Ma T, Pesenti S, Sonne SB, Marstrand TT, Malde MK, Du ZY, Chavey C, Fajas L, Lundebye AK, Brand CL, Vidal H, Kristiansen K, Frøyland L (2010) Persistent organic pollutant exposure leads to insulin resistance syndrome. Environ Health Perspect 118:465–471

Seegal RF, Fitzgerald EF, Hills EA, Wolff MS, Haase RF, Todd AC, Parsons P, Molho ES, Higgins DS, Factor SA, Marek KL, Seibyl JP, Jennings DL, McCaffrey RJ (2010) Estimating the half-lives of PCB congeners in former capacitor workers measured over a 28-year interval. J Expo Sci Environ Epidemiol. doi:10.1038/jes.2010.3

Turyk M, Anderson H, Knobeloch L, Imm P, Persky V (2009) Organochlorine exposure and incidence of diabetes in a cohort of great lakes sport fish consumers. Environ Health Perspect 117:1076–1082

Ukropec J, Radikova Z, Huckova M, Koska J, Kocan A, Sebokova E, Drobna B, Trnovec T, Susienkova K, Labudova V, Gasperikova D, Langer P, Klimes I (2010) High prevalence of prediabetes and diabetes in a population exposed to high levels of an organochlorine cocktail. Diabetologia 53:899–906

Wang SL, Tsai PC, Yang CY, Leon Guo Y (2008) Increased risk of diabetes and polychlorinated biphenyls and dioxins: a 24-year follow-up study of the Yucheng cohort. Diabetes Care 31:1574–1579

Weihe P, Kato K, Calafat AM, Nielsen F, Wanigatunga AA, Needham LL, Grandjean P (2008) Serum concentrations of polyfluoroalkyl compounds in Faroese whale meat consumers. Environ Sci Technol 42:6291–6295

Wolff MS, Anderson HA, Britton JA, Rothman N (2007) Pharmacokinetic variability and modern epidemiology – the example of dichlorodiphenyltrichloroethane, body mass index, and birth cohort. Cancer Epidemiol Biomarkers Prev 16:1925–1930

World Health Organization (WHO) (2010) Diabetes programme 2010. World Health Organization http://www.who.int./diabetes.en/. Accessed July 22 2011.

Contribution of Endocrine Disrupting Chemicals to the Obesity Epidemic: Consequences of Developmental Exposure

Retha R. Newbold

Abstract Some environmental chemicals are known to disrupt the programming of endocrine signaling pathways that are established during development, resulting in adverse effects later in life. Initially, most endocrine disruptor studies focused on alterations in fertility and reproductive tract endpoints; however, recent evidence implicates developmental exposure to endocrine disrupting chemicals (EDCs) with a growing list of adverse health consequences, including an association with obesity and diabetes. These diseases and their related complications are quickly becoming significant public health problems worldwide and are fast reaching epidemic proportions in many countries. Herein, data from experimental animals are summarized that show an association of environmental estrogens – such as the synthetic estrogen diethylstilbestrol (DES), the high-volume production monomer bisphenol A (BPA) used in the production of polycarbonate plastics, and phytoestrogens found in foods such as soy products – with the development of obesity. In addition, the link to other EDCs with various hormone-disruption activities, such as organotins, phthalates, pesticides, and persistent organic pollutants, are discussed. These animal studies are supported by both experiments with cells in culture and epidemiology surveys that suggest that EDCs contribute to the obesity/diabetes epidemic. The association of EDCs with obesity and related diseases does not diminish the roles of diet and exercise; however, it does point out that exposures to EDCs during development are risk factors that should be avoided wherever possible. This idea shifts the focus on obesity from treatment to prevention.

The author is retired but the research was conducted while employed by NIEHS.

R.R. Newbold (✉)
National Institute of Environmental Health Sciences (NIEHS), National Institutes of Health (NIH), Department of Health and Human Services (DHHS), Durham, NC, USA
e-mail: newbold1@niehs.nih.gov

J.-P. Bourguignon et al. (eds.), *Multi-System Endocrine Disruption*, Research and Perspectives in Endocrine Interactions,
DOI 10.1007/978-3-642-22775-2_7, © Springer-Verlag Berlin Heidelberg 2011

Introduction

Obesity, defined as a body mass index (BMI) greater than 30, has been identified as a significant public health threat (Oken and Gillman 2003; Ogden et al. 2007). The prevalence of obesity over the last two to three decades has risen dramatically in wealthy industrialized countries and also in poorer underdeveloped nations, where it often coexists with under-nutrition (Caballero 2007; Cunningham 2010). In 2008, the Center for Disease Control (CDC) in the United States reported that obesity had reached alarming epidemic proportions, with more than 60% of U.S. adults being either obese or overweight. Further, a more recent report estimates that approximately 40% of people who are 20 years of age and older in the U.S. have diabetes or pre-diabetes (Cowie et al. 2009), and there is a continuing increase in age-adjusted prevalence in metabolic syndrome (Mozumdar and Liguori 2011). Similar statistics have been reported for many European countries, the Middle East, Australia and China. The reasons for this sharp increase in obesity and diabetes are not known but multiple factors (infection, high fructose diets, genetics/epigenetics, increased maternal age, sleep deficient, stress, use of certain pharmaceuticals, the built environment, etc.) have been proposed to play a role (McAllister et al. 2009).

Obesity is a problem for all ages but it is of particular concern for children, since most obese and overweight children grow up to become obese adults, predisposing them to a lifetime of adverse health consequences. Unfortunately, the number of children and adolescents who are overweight, or at risk for being overweight, has risen in parallel with that reported in adults (Ogden et al. 2002). Although one recent study suggests that the increase in the prevalence of obesity may have finally leveled off in the last few years, there is still no indication of any decreases occurring (Flegal et al. 2010).

Obesity is a complex disease, affecting virtually all ages, races, sexes, and socioeconomic groups, and it has serious social and psychological repercussions. The exact etiology of obesity is unknown but it is most likely caused by a complicated interaction between genetic, behavioral, and environmental factors. Traditional risk factors for obesity are thought to involve overeating high caloric fatty diets combined with a sedentary lifestyle. Thus, justifiably, much interest has centered on these factors, with a specific goal of incorporating healthy foods in our diets and more exercise in our lifestyle. Yet treatment and prevention of obesity remain elusive. Since obesity is such a challenge to treat effectively once it is established, it is important to focus on non-traditional risk factors that may help in the prevention of this disease.

For many years, adipocytes were considered just inert energy storage "blobs." However, the discovery of leptin in 1994 greatly transformed ideas about "fat" cells. It became quickly apparent that adipocytes were active endocrine tissues, secreting leptin and numerous other factors and playing an essential role in the endocrine feedback loop with the CNS, stomach, liver, muscle, pancreas, thyroid, etc., within the body to regulate appetite, satiety, and energy expenditure. Many other metabolically active factors produced by adipocytes were subsequently identified following leptin, including adiponectin, visfatin, resistin, interleukin-6 (IL-6), and retinol-binding protein-4 (RBP-4) (Isganaitis and Patti 2011), which firmly established fat cells as active endocrine tissues. Interestingly, many of the adipocyte-signaling pathways were determined to be established during prenatal and neonatal development. Thus, any disruption in these signaling pathways (for example, leptin signaling) could have long-lasting consequences.

Recent evidence points to the contribution of EDCs in the increasing prevalence of obesity (Baillie-Hamilton 2002; Heindel 2003; Heindel and Levin 2005; Newbold et al. 2005, 2007a, b, 2008). EDCs are chemicals, synthetic or naturally occurring, that mimic or interfere with the production or activity of hormones of the endocrine system, leading to adverse health effects. Exposure to EDCs, especially during prenatal or postnatal development, can interfere with the normal function of the endocrine system by affecting the balance of hormones/signals that regulate vital body functions, including growth, stress response, sex development, gender behavior, ability to reproduce, production and utilization of insulin, and metabolic rate. EDCs that are specifically associated with obesity are referred to as "obesogens," referring to the idea that they may inappropriately regulate lipid metabolism and/or adipogenesis to promote obesity (Grun and Blumberg 2006; Grun et al. 2006). Diethylstilbestrol (DES) is an example of an EDC with estrogenic activity. It provides the proof of principle that developmental exposure to EDCs is a risk factor for obesity later in life.

Exposure to Environmental Obesogenic Chemicals

Diethylstilbestrol (DES)

DES, a potent synthetic non-steroidal estrogen, was prescribed for millions of pregnant women from the 1940s through the 1970s with the mistaken belief that it could prevent threatened miscarriages. Today, it is well documented that prenatal DES treatment resulted in a low but significant increase in neoplastic lesions in the reproductive tract and a high incidence of benign reproductive lesions in both the male and female offspring exposed during prenatal life. Administration of DES to pregnant mice causes similar abnormalities and dysfunction in the reproductive tract of their offspring (Newbold 2004).

Developmental (either prenatal or neonatal) exposure to DES over a wide dose range of exposures also caused the murine offspring to gain excessive weight and have a high BMI (Newbold 2011); for example, at 6 weeks of age, control mice were estimated to have 6.5 ± 0.8 vs. 11.4 ± 1.3 g fat weight compared to DES-treated (0.001 mg/kg) mice, as determined by PIXImus mouse densitometry (Newbold et al. 2005). However, DES did not alter their food intake or activity levels (Newbold et al. 2007b); this finding implies a change in metabolic regulation of energy. Serum profiles from these DES-treated animals showed increased levels of leptin, adiponectin, and IL-6 compared to age matched controls at 2 months of age even before obesity was apparent; by 6 months of age, the DES-mice became obviously obese compared to controls and all serum endpoints remained elevated. Insulin was also shown to be elevated by this age (Newbold et al. 2007a, b). DES-exposed mice had enhanced abdominal circumferences which, when reported in humans, is considered to be a risk factor for cardiovascular disease associated with obesity and metabolic syndrome.

To determine if DES was a unique estrogen in causing increased weight gain, mice were treated neonatally with other estrogenic compounds, such as 2OH estradiol (20 mg/kg/day) or 4 OH estradiol (0.1 mg/kg/day), using doses that are approximately equal in estrogenic activity to the low dose DES (0.001 mg/kg). These two compounds also caused a significant increase in body weight, indicating that DES is not the only estrogen capable of causing overweight and obesity (Newbold et al. 2005).

Glucose levels were measured in DES (1,000 μg/kg/day = 1 mg/kg/day) and control mice at 2 months of age prior to the development of obesity and excessive weight gain (Newbold et al. 2007b). Interestingly, 25% of the DES-treated mice had significantly higher glucose levels than controls; these DES mice also showed a slower clearance rate of glucose from the blood, since higher levels were seen throughout the experiment (Newbold et al. 2007b). Additional glucose measurements in older mice may help determine if higher percentages of mice are affected with age and if higher and sustained levels of glucose can be demonstrated. To date, however, our data suggest that overweight and obesity observed in perinatal DES-treated mice will be associated with the development of diabetes, similar to the association of obesity with diabetes in humans. Further, other studies from our laboratory support a role for altered glucose metabolism, since we have shown a high prevalence of islet cell hyperplasia in the pancreas of mice exposed to DES or other environmental estrogens (unpublished data).

When a study described a role for developmental genes in the origins of obesity and body fat distribution in mice and humans (Gesta et al. 2006), we determined whether exposure to environmental chemicals with hormonal activity altered the gene expression involved in programming adipocytes during development. Several genes were implicated in altered adipocyte differentiation and function (*Hoxa5*, *Gpc4* and *Tbx15)* and fat cell distribution (*Thbd*, *Nr2f1* and *Sfrp2)*. We investigated changes in gene expression by microarray analysis in uterine samples from DES-treated mice (1,000 μg/kg/day = 1 mg/kg/day) compared to controls at 19 days of age. Genes involved in adipocyte differentiation were not different in the uterus following neonatal DES exposure. However, genes involved in fat distribution were altered; *Thbd* and *Nr2f1* were significantly down regulated and *Sfrp2* was significantly up regulated in DES-treated uteri compared to controls (Newbold et al. 2007c). These findings support the idea that environmental estrogens may play a role in regulating the expression of obesity-related genes in development. The identification of genes and molecular mechanisms that may be associated with EDCs and obesity is a promising area of new research.

Other investigators have also reported similar findings of an association of perinatal DES exposure with the development of obesity later in life (Nikaido et al. 2004). Nikaido and colleagues treated pregnant mice with DES beginning on gestational days 15–19 with 0.5 or 10 μg/kg. By 16 weeks of age, all the female DES offspring were heavier than corresponding controls, although the lowest dose was not significantly different.

Although DES is no longer used for treatment during pregnancy, it is an excellent experimental prototype chemical, providing essential information about the mechanisms of environmental EDCs with estrogenic activity, especially when low does of DES are studied. Since steroid hormones and their receptors have a long history with growth and metabolism, the role of chemicals with hormone-like activity should be interesting to pursue.

Bisphenol A (BPA)

BPA, a component of polycarbonate plastics and epoxy resins, is currently receiving much attention from the public, governmental regulatory agencies, and industry advocates, due to its high production volume (>800 million kilograms annually in the U.S. alone) and widespread human exposure (Zoeller et al. 2005). BPA is used in the manufacture of numerous products and has been shown to leach from the linings of food cans (Brotons et al. 1994), polycarbonate baby bottles and other beverage containers (Biles et al. 1997), and dental sealants and composites (Olea et al. 1996), suggesting that humans are routinely exposed to this chemical through numerous sources and routes of exposure. Further indication of human exposure is shown by studies reporting measurable BPA levels in human urine (Calafat et al. 2008), serum (Takeuchi and Tsutsumi 2002), breast milk (Ye et al. 2006), and maternal and fetal plasma, amniotic fluid, and placental tissues (Schonfelder et al. 2002; Padmanabhan et al. 2008). BPA is also halogenated (brominated or chlorinated) to produce flame retardants; tetrabromobisphenol A (TBBPA) is the most commonly used, with >60,000 tons produced annually (WHO 1995; European Union Risk Assessment Report 2008). Studies report that levels of brominated flame retardants are increasing in the serum of adults; most importantly, the level in infants and children is markedly higher than in adults (Thomsen et al. 2002).

BPA is often described as a "weak" estrogen; however, an emerging number of cellular and molecular studies find that it has potential for many other biological activities at low, environmentally relevant exposure levels. In addition to binding to the nuclear estrogen receptor (ER) alpha and ER beta, BPA interacts with a variety of other cellular targets, including binding to a non-classical membrane-bound form of the ER (ncmER), a recently identified orphan nuclear receptor termed estrogen-related receptor gamma (ERR gamma), a seven trans-membrane estrogen receptor called GPR30, and the aryl hydrocarbon receptor (AhR). Interactions with ncmER and ERR gamma are especially noteworthy because BPA binds to these receptors with high affinity. BPA has also been shown to act as an androgen receptor antagonist and to interact with thyroid hormone receptors (for review, see Wetherill et al. 2007).

Experimental animal studies have reported that very low doses of BPA in the range of human exposures can exert effects if given during development. An increasing number of "low dose" studies has suggested perinatal BPA exposure is associated with a variety of abnormalities in the male and female reproductive tissues and mammary gland tissues (for review of low dose BPA effects, see Richter et al. 2007 and NTP 2008). Studies showing an association between BPA and obesity have received less attention; however, data from both mice and rats have shown increased body weights in animals that were exposed to low doses of BPA during prenatal or neonatal development (Ashby et al. 1999; Howdeshell et al. 1999; Rubin et al. 2001; Takai et al. 2001; Honma et al. 2002; Nikaido et al. 2004, 2005; Somm et al. 2009). The latter study suggests this increase in body weight is sex specific, but timing and dose may also contribute to the complexity of these findings since other studies reported effects in both males and females. Interestingly, a recent paper reports similar increases in body weights of pups obtained from moms fed BPA in their diets during pregnancy; the doses were low and considered "ecologically relevant" at 1 μg/kg diet (1 ppb) BPA (Ryan et al. 2010). So regardless of the route of exposure, low doses of BPA caused weight gain in pups. Unlike our results with BPA and the findings of other investigators, Ryan and colleagues report that the differences seen in body weight at weaning disappear as the mice age (Ryan et al. 2010). This finding is probably due to the palatability of the diet, which was changed at weaning, since neither control nor BPA mice continued to gain weight on the new diets. Taken together, these experimental studies on BPA point out the sensitivity of the developmental period to disruption by environmental chemicals, but they also suggest the complexity of the mechanisms involved in the development of obesity.

In vitro studies with BPA provide additional evidence for a role for the chemical in the development of obesity and further suggest specific targets; BPA causes 3T3-L1 cells (mouse fibroblast cells that can differentiate into adipocytes) to increase differentiation (Sakurai et al. 2004) and, in combination with insulin, BPA accelerates adipocyte formation (Masuno et al. 2002, 2005). Other in vitro studies also show that low doses of BPA, similar to DES, impair calcium signaling in pancreatic alpha cells, disrupt beta cell function, and cause insulin resistance (Alonso-Magdalena et al. 2005, 2006). Low, environmentally relevant doses of BPA have also been reported to inhibit adiponectin and stimulate the release of inflammatory adipokines such as interleukin-6 (IL-6) and tumor necrosis factor alpha (TNF-α) from human adipose tissue, suggesting that BPA is involved in obesity and the related metabolic syndrome (Hugo et al. 2008; Ben-Jonathan et al. 2009). Further, other studies have linked BPA exposure to disruption of pancreatic beta-cell function and blood glucose homeostasis in mice (Ropero et al. 2008), suggesting changes indicative of metabolic syndrome.

Epidemiology studies also support an association between BPA and obesity. BPA was detected at high levels in both non-obese and obese women with polycystic ovarian syndrome (PCOS), compared with BPA levels in non-obese normal women, suggesting the possible involvement of BPA in PCOS and/or obesity (Takeuchi et al. 2004).

Phytoestrogens

In recent years, efforts to implement healthier eating lifestyles have resulted in an increased consumption of soy products and soy supplements, which in turn caused increased exposure to phytoestrogens. Although phytoestrogens are being considered as a treatment or a preventative for obesity, they may have a role in causing the disease. Genistein and daidzein are two of the most abundant phytoestrogens in the human diet; genistein, because of its estrogenic activity, has been proposed to have a role in the maintenance of health by regulating lipid and carbohydrate homeostasis. However, a recent study showed that, although genistein at pharmacologically high doses inhibited adipose deposition, at low doses similar to that found in Western and Eastern diets, in soy milk, or in food supplements containing soy, it induced adipose tissue deposition especially in males (Penza et al. 2006). Further, this increase in adipose tissue deposition by genistein was correlated with mild peripheral insulin resistance. Similar to our findings with DES, genistein caused abnormal programming of factors involved in weight homeostasis but it did not affect food consumption (Penza et al. 2006). Certainly, additional research is needed but the data thus far suggest that genistein can affect adipose tissue deposition and that the effects are dose-dependent and gender-specific.

Epidemiologic studies also support the idea that phytoestrogens may contribute to overweight babies if they are fed soy-based infant formula (Stettler et al. 2005; for a review of the effects of soy formula, see Rozman et al. 2006).

Developmental exposure to numerous other naturally estrogenic substances has been studied in animals and their effects on body weight well documented, including increased body weight caused by resveratrol found in grapes and red wine, and zearalenone, a mycotoxin synthesized by Fusarium mold and present in "moldy" grains, especially corn (Nikaido et al. 2004). Phytoestrogens and mycoestrogens, both singly or in combination with other chemicals, may have multiple pathways of action and are likely involved in adipogenesis and energy metabolism. Whether they have any beneficial, harmful, or no effects remains to be determined, but it most likely depends on timing of exposure and the dose.

Tributyltin

Unlike the chemicals with estrogenic activity that have been discussed, there is another class of chemicals that has been linked to the development of obesity (Janesick and Blumberg 2011). Organotins are chemical compounds prevalent in industry and used in fungicides, wood preservatives, and heat stabilizers in some plastics. They were also widely used as antifouling paints on boats; since their regulation, their use has decreased but has not been eliminated. Since organotins are lipophilic, they can readily bioaccumulate. Concern over exposure to organotins arose when female mollusks exposed to tributyltin (TBT) were reported to be masculinized; additional reproductive effects have since been reported in other species. In addition, it has subsequently been found that *Xenopus laevis* tadpoles exposed to low levels of TBT had ectopic fat production. In mice, a single prenatal exposure to TBT during gestation resulted in premature accumulation of fat in adipose tissues (Grun et al. 2006). Although the mice were born slightly underweight, they already had stored fat at the expense of total body mass. By 6 months of age, the TBT mice were significantly heavier than age-matched controls. Similar to our observations with DES-treated mice, the TBT mice showed no increase in food intake or decrease in exercise compared to controls. Histological sections of newborn liver, testis, mammary gland, and inguinal adipose tissue (which do not usually store fat before eating commences) all showed pronounced lipid accumulation in the pups born to TBT-treated moms. The tendency to store excess fat was already programmed before the mouse was born, due to one treatment with TBT to the mom, at a low dose similar to human exposure. This obese phenotype was seen in both males and females.

Other Chemicals

An overview of the toxicology literature suggests that developmental exposure to many other environmental chemicals can cause weight gain: for example, dioxins, DDT, endrin, lindane, and hexachlorobenzene; organophosphates; carbamates; polychlorinated biphenyls; polybrominated biphenols, which are used as fire retardants; other plastic components such as phthalates; perfluoroctanoic acid (PFOA); heavy metals such as cadmium, lead and arsenic; and solvents (for review, see Baillie-Hamilton 2002). The weight gain associated with these chemicals tends to occur at low levels of exposure not at high doses, where most toxicity studies have been conducted. Further, it is well known that many pharmaceuticals also play a role in altered weight homeostasis and metabolism and/or altered hormone levels; in fact, many prescribed drugs have the unwanted side effects of weight gain, including drugs commonly used in oncology, cardiology, immunology and psychiatry (Baillie-Hamilton 2002). While these medicines are normally prescribed for adults where their effects on weight gain would most likely be reversible and disappear when medication ceased, the possible impact on the unborn fetus and young child should not be underestimated due to inadvertent contamination. With developmental exposure, the effects would likely be permanent and non-reversible. The wide group of chemicals that have been associated with obesity suggests that multiple complex mechanisms are most likely involved.

Summary and Conclusions

The data included in this review support the idea that brief exposure, early in development, to environmental chemicals with endocrine disruptor activity increases body weight gain with age and alters markers predictive of obesity in experimental animals. Further, both epidemiology and in vitro studies support the findings in experimental animals and show a link between exposure to EDCs and the development of obesity (Baillie-Hamilton 2002; Heindel 2003). Using the DES animal model as an important research tool to study "obesogens," the mechanisms involved in altered weight homeostasis (direct and/or endocrine feedback loops, i.e., ghrelin, leptin, etc.) by environmental estrogens and other EDCs can be elucidated (Newbold et al. 2005, 2007a, b, 2008). In addition, this new focus on environmental chemicals and obesity may shed light on areas of prevention. Public health risks can no longer be based on the assumption that overweight and obesity are just personal choices involving the quantity and kind of foods we eat combined with inactivity; rather, complex events including exposure to environmental chemicals during development may be contributing to the obesity epidemic.

References

Alonso-Magdalena P, Laribi O, Ropero AB, Fuentes E, Ripoll C, Soria B, Nadal A (2005) Low doses of bisphenol A and diethylstilbestrol impair Ca^{2+} signals in pancreatic alpha-cells through a nonclassical membrane estrogen receptor within intact islets of Langerhans. Environ Health Perspect 113:969–977

Alonso-Magdalena P, Morimoto S, Ripoll C, Fuentes E, Nadal A (2006) The estrogenic effect of bisphenol A disrupts pancreatic beta-cell function in vivo and induces insulin resistance. Environ Health Perspect 114:106–112

Ashby J, Tinwell H, Haseman J (1999) Lack of effects for low dose levels of bisphenol A and diethylstilbestrol on the prostate gland of CF1 mice exposed in utero. Regul Toxicol Pharmacol 30:156–166

Baillie-Hamilton PF (2002) Chemical toxins: a hypothesis to explain the global obesity epidemic. J Altern Complement Med 8:185–192

Ben-Jonathan N, Hugo ER, Brandebourg TD (2009) Effects of bisphenol A on adipokine release from human adipose tissue: implications for the metabolic syndrome. Mol Cell Endocrinol 304:49–54

Biles JE, McNeal TP, Begley TH, Hollifield HC (1997) Determination of bisphenol-A in reusable polycarbonate food-contact plastics and migration to food simulating liquids. J Agric Food Chem 45:3541–3544

Brotons JA, Olea-Serrano MF, Villalobos M, Olea N (1994) Xenoestrogens released from lacquer coating in food cans. Environ Health Perspect 103:608–612

Caballero B (2007) The global epidemic of obesity: an overview. Epidemiol Rev 29:1–5

Calafat AM, Ye X, Wong LY, Reidy JA, Needham LL (2008) Exposure of the U.S. population to bisphenol A and 4-tertiary-octylphenol: 2003–2004. Environ Health Perspect 116:39–44

CDC (2008) Report on overweight and obesity. Centers for Disease Control and Prevention. http://www.cdc.gov/nccdphp/dnpa/obesity2011

Cowie CC, Rust KF, Ford ES, Eberhardt MS, Byrd-Holt DD, Li C, Williams DE, Gregg EW, Bainbridge KE, Saydah SH, Geiss LS (2009) Full accounting of diabetes and pre-diabetes in the U.S. population in 1988–1994 and 2005–2006. Diabetes Care 32:287–294

Cunningham E (2010) Where can I find obesity statistics? J Am Diet Assoc 110:656

European Union Updated European Risk Assessment Report (2008) 4,4′-isopropylidenediphenol (bisphenol A). Environment addendum of February 2008 (to be read in conjunction with EU RAR of BPA published in 2003). http:ecb.jrc.it/documents/Existing-Chemicals/RISKASSESSMENT/ADDENDUM/bisphenola add 325.pdf2011

Flegal KM, Carroll MD, Ogden CL, Curtin LR (2010) Prevalence and trends in obesity among US adults, 1999–2008. JAMA 303:235–241

Gesta S, Bluher M, Yamamoto Y, Norris AW, Berndt J, Kralisch S, Boucher J, Lewis C, Kahn CR (2006) Evidence for a role of developmental genes in the origin of obesity and body fat distribution. Proc Natl Acad Sci USA 103:6676–6681

Grun F, Blumberg B (2006) Environmental obesogens: organotins and endocrine disruption via nuclear receptor signaling. Endocrinology 147:S50–S55

Grun F, Watanabe H, Zamanian Z, Maeda L, Arima K, Cubacha R, Gardiner DM, Kanno J, Iguchi T, Blumberg B (2006) Endocrine-disrupting organotin compounds are potent inducers of adipogenesis in vertebrates. Mol Endocrinol 20:2141–2155

Heindel JJ (2003) Endocrine disruptors and the obesity epidemic. Toxicol Sci 76:247–249

Heindel JJ, Levin E (2005) Developmental origins and environmental influences – introduction. NIEHS symposium. Birth Defect Res A Clin Mol Teratol 73:469

Honma S, Suzuki A, Buchanan DL, Katsu Y, Watanabe H, Iguchi T (2002) Low dose effect of in utero exposure to bisphenol A and diethylstilbestrol on female mouse reproduction. Reprod Toxicol 16:117–122

Howdeshell KL, Hotchkiss AK, Thayer KA, Vandenbergh JG, vom Saal FS (1999) Exposure to bisphenol A advances puberty. Nature 401:763–764

Hugo ER, Brandebourg TD, Woo JG, Loftus J, Alexander JW, Ben-Jonathan N (2008) Bisphenol A at environmentally relevant doses inhibits adiponectin release from human adipose tissue explants and adipocytes. Environ Health Perspect 116:1642–1647

Isganaitis E, Patti M (2011) Adipocyte development and experimental obesity. In: Lustig RH (ed) Obesity before birth. Springer, New York, pp 321–352

Janesick A, Blumberg B (2011) The role of environmental obesogens in the obesity epidemic. In: Lustig RH (ed) Obesity before birth. Springer, New York, pp 383–399

Masuno H, Kidani T, Sekiya K, Sakayama K, Shiosaka T, Yamamoto H, Honda K (2002) Bisphenol A in combination with insulin can accelerate the conversion of 3T3-L1 fibroblasts to adipocytes. J Lipid Res 43:676–684

Masuno H, Iwanami J, Kidani T, Sakayama K, Honda K (2005) Bisphenol a accelerates terminal differentiation of 3T3-L1 cells into adipocytes through the phosphatidylinositol 3-kinase pathway. Toxicol Sci 84:319–327

McAllister EJ, Dhurandhar NV, Keith SW, Aronne LJ, Barger J, Baskin M, Benca RM, Biggio J, Boggiano MM, Eisenmann JC, Elobeid M, Fontaine KR, Gluckman P, Hanlon EC, Katzmarzyk P, Pietrobelli A, Redden DT, Ruden DM, Wang C, Waterland RA, Wright SM, Allison DB (2009) Ten putative contributors to the obesity epidemic. Crit Rev Food Sci Nutr 49:868–913

Mozumdar A, Liguori G (2011) Persistent increase of prevalence of metabolic syndrome among U.S. adults: NHANES III to NHANES 1999–2006. Diabetes Care 34:216–219

Newbold R (2004) Lessons learned from perinatal exposure to diethylstilbestrol (DES). Toxicol Appl Pharmacol 199:142–150

Newbold R (2011) Perinatal exposure to endocrine disrupting chemicals with estrogenic activity and the development of obesity. In: Lustig RH (ed) Obesity before birth. Springer, New York, pp 367–382

Newbold RR, Padilla-Banks E, Snyder RJ, Jefferson WN (2005) Developmental exposure to estrogenic compounds and obesity. Birth Defect Res A Clin Mol Teratol 73:478–480

Newbold RR, Padilla-Banks E, Snyder RJ, Jefferson WN (2007a) Perinatal exposure to environmental estrogens and the development of obesity. Mol Nutr Food Res 51:912–917
Newbold RR, Padilla-Banks E, Snyder RJ, Phillips TM, Jefferson WN (2007b) Developmental exposure to endocrine disruptors and the obesity epidemic. Reprod Toxicol 23:290–296
Newbold RR, Jefferson WN, Grissom SF, Padilla-Banks E, Snyder RJ, Lobenhofer EK (2007c) Developmental exposure to diethylstilbestrol alters uterine gene expression that may be associated with uterine neoplasia later in life. Mol Carcinog 46:783–796
Newbold RR, Padilla-Banks E, Jefferson WN, Heindel JJ (2008) Effects of endocrine disruptors on obesity. Int J Androl 31:201–208
Nikaido Y, Yoshizawa K, Danbara N, Tsujita-Kyutoku M, Yuri T, Uehara N, Tsubura A (2004) Effects of maternal xenoestrogen exposure on development of the reproductive tract and mammary gland in female CD-1 mouse offspring. Reprod Toxicol 18:803–811
Nikaido Y, Danbara N, Tsujita-Kyutoku M, Yuri T, Uehara N, Tsubura A (2005) Effects of prepubertal exposure to xenoestrogen on development of estrogen target organs in female CD-1 mice. In Vivo 19:487–494
NTP (2008) CEHR brief on bisphenol A. National Toxicology Program. Research Triangle Park, NC
Ogden CL, Flegal KM, Carroll MD, Johnson CL (2002) Prevalence and trends in overweight among US children and adolescents, 1999–2000. JAMA 288:1728–1732
Ogden CL, Yanovski SZ, Carroll MD, Flegal KM (2007) The epidemiology of obesity. Gastroenterology 132:2087–2102
Oken E, Gillman MW (2003) Fetal origins of obesity. Obes Res 11:496–506
Olea N, Pulgar R, Perez P, Olea-Serrano F, Rivas A, Novillo-Fertrell A, Pedraza V, Soto AM, Sonnenschein C (1996) Estrogenicity of resin-based composites and sealants used in dentistry. Environ Health Perspect 104:298–305
Padmanabhan V, Siefert K, Ransom S, Johnson T, Pinkerton J, Anderson L, Tao L, Kannan K (2008) Maternal bisphenol-A levels at delivery: a looming problem? J Perinatol 28:258–263
Penza M, Montani C, Romani A, Vignolini P, Pampaloni B, Tanini A, Brandi ML, Alonso-Magdalena P, Nadal A, Ottobrini L, Parolini O, Bignotti E, Calza S, Maggi A, Grigolato PG, Di Lorenzo D (2006) Genistein affects adipose tissue deposition in a dose-dependent and gender-specific manner. Endocrinology 147:5740–5751
Richter CA, Birnbaum LS, Farabollini F, Newbold RR, Rubin BS, Talsness CE, Vandenbergh JG, Walser-Kuntz DR, vom Saal FS (2007) In vivo effects of bisphenol A in laboratory rodent studies. Reprod Toxicol 24:199–224
Ropero AB, Alonso-Magdalena P, Garcia-Garcia E, Ripoll C, Fuentes E, Nadal A (2008) Bisphenol-A disruption of the endocrine pancreas and blood glucose homeostasis. Int J Androl 31:194–200
Rozman KK, Bhatia J, Calafat AM, Chambers C, Culty M, Etzel RA, Flaws JA, Hansen DK, Hoyer PB, Jeffery EH, Kesner JS, Marty S, Thomas JA, Umbach D (2006) NTP-CERHR expert panel report on the reproductive and developmental toxicity of soy formula. Birth Defect Res B Dev Reprod Toxicol 77:280–397
Rubin BS, Murray MK, Damassa DA, King JC, Soto AM (2001) Perinatal exposure to low doses of bisphenol A affects body weight, patterns of estrous cyclicity, and plasma LH levels. Environ Health Perspect 109:675–680
Ryan KK, Haller AM, Sorrell JE, Woods SC, Jandacek RJ, Seeley RJ (2010) Perinatal exposure to bisphenol-A and the development of metabolic syndrome in CD-1 mice. Endocrinology 151:2603–2612
Sakurai K, Kawazuma M, Adachi T, Harigaya T, Saito Y, Hashimoto N, Mori C (2004) Bisphenol A affects glucose transport in mouse 3T3-F442A adipocytes. Br J Pharmacol 141:209–214
Schonfelder G, Flick B, Mayr E, Talsness C, Paul M, Chahoud I (2002) In utero exposure to low doses of bisphenol A lead to long-term deleterious effects in the vagina. Neoplasia 4:98–102

Somm E, Schwitzgebel VM, Toulotte A, Cederroth CR, Combescure C, Nef S, Aubert ML, Huppi PS (2009) Perinatal exposure to bisphenol a alters early adipogenesis in the rat. Environ Health Perspect 117:1549–1555

Stettler N, Stallings VA, Troxel AB, Zhao J, Schinnar R, Nelson SE, Ziegler EE, Strom BL (2005) Weight gain in the first week of life and overweight in adulthood: a cohort study of European American subjects fed infant formula. Circulation 111:1897–1903

Takai Y, Tsutsumi O, Ikezuki Y, Kamei Y, Osuga Y, Yano T, Taketan Y (2001) Preimplantation exposure to bisphenol A advances postnatal development. Reprod Toxicol 15:71–74

Takeuchi T, Tsutsumi O (2002) Serum bisphenol a concentrations showed gender differences, possibly linked to androgen levels. Biochem Biophys Res Commun 291:76–78

Takeuchi T, Tsutsumi O, Ikezuki Y, Takai Y, Taketani Y (2004) Positive relationship between androgen and the endocrine disruptor, bisphenol A, in normal women and women with ovarian dysfunction. Endocr J 51:165–169

Thomsen C, Lundanes E, Becher G (2002) Brominated flame retardants in archived serum samples from Norway: a study on temporal trends and the role of age. Environ Sci Technol 36:1414–1418

Wetherill YB, Akingbemi BT, Kanno J, McLachlan JA, Nadal A, Sonnenschein C, Watson CS, Zoeller RT, Belcher SM (2007) In vitro molecular mechanisms of bisphenol A action. Reprod Toxicol 24:178–198

WHO (1995) Tetrabromobisphenol A and derivatives, vol 172, Environmental health criteria. World Health Organization, Geneva

Ye X, Kuklenyik Z, Needham LL, Calafat AM (2006) Measuring environmental phenols and chlorinated organic chemicals in breast milk using automated on-line column-switching-high performance liquid chromatography-isotope dilution tandem mass spectrometry. J Chromatogr B Analyt Technol Biomed Life Sci 831:110–115

Zoeller RT, Bansal R, Parris C (2005) Bisphenol-A, an environmental contaminant that acts as a thyroid hormone receptor antagonist in vitro, increases serum thyroxine, and alters RC3/neurogranin expression in the developing rat brain. Endocrinology 146:607–612

Fetal and Adult Exposure to Bisphenol-A as a Contributing Factor in the Etiology of the Metabolic Syndrome

Paloma Alonso-Magdalena and Angel Nadal

Abstract Metabolic disorders such as type 2 diabetes and obesity are among the most challenging health problems on a global scale. The number of patients is increasing worldwide at an alarming rate. Although the underlying cause of the problem is still puzzling, genetic and environmental factors are thought to have a causal influence.

Furthermore, widespread human exposure to significant doses of bisphenol-A (BPA) has been reported. BPA is a product commonly used in food and beverage containers that has been demonstrated to interfere with endocrine signaling pathways at low doses during fetal, neonatal or perinatal periods as well as in adulthood. There is also increasing experimental evidence revealing the deleterious effects of BPA on energy balance and glucose homeostasis.

In the present review, we will summarize the most relevant findings that confirm the critical role of BPA in the development of metabolic disorders.

Endocrine and metabolic diseases remain the leading causes of morbidity and mortality in the modern world. Type 2 diabetes, obesity and associated cardiovascular diseases are on the rise most likely due to unhealthy eating patterns along with sedentary lifestyles. Traditionally, these conditions were considered to be confined to the middle-aged and elderly, but the emerging incidence of obesity-associated type 2 diabetes has been accompanied by its appearance in adolescents and children (Dabelea et al. 2010; Fox et al. 2006; Rosenbloom et al. 1999). Although, there is a strong genetic component in the susceptibility to these diseases, their rapidly increasing incidence seems difficult to explain just as a result of genetic changes,

P. Alonso-Magdalena (✉)
Instituto de Bioingeniería and CIBERDEM, Universidad Miguel Hernández de Elche, Elche, Spain
e-mail: palonso@umh.es

J.-P. Bourguignon et al. (eds.), *Multi-System Endocrine Disruption*, Research and Perspectives in Endocrine Interactions,
DOI 10.1007/978-3-642-22775-2_8, © Springer-Verlag Berlin Heidelberg 2011

thus emphasizing the importance of environmental factors (O'Rahilly 2009; Qatanani and Lazar 2007).

Moreover, we have learned throughout time that many chemical substances can cause a range of adverse health problems, including neonatal defects, cancer, and a delay in the development of cognitive functions. For instance, it is well established that thalidomide can cause limb deformities (Ito et al. 2010), asbestos can cause a fatal form of lung cancer (Heintz et al. 2010), and breathing high concentrations of some industrial solvents can provoke irreversible brain damage and death (Grandjean and Landrigan 2006). We have only recently become aware that a large number of chemicals, collectively termed endocrine disrupting chemicals (EDCs), show estrogenic activity and that they can interfere with complex endocrine signalling pathways, causing adverse consequences in a developing organism (Bern 1992; Colborn et al. 1996; Newbold et al. 2009). Interestingly, a new area of study has arisen around the implication of EDCs in the etiology of obesity and diabetes (Heindel and vom Saal 2009; Newbold et al. 2008; Ropero et al. 2008a; vom Saal and Myers 2008).

In this regard, special attention has been paid to the compound bisphenol-A (BPA), a chemical used in plastics and widely spread in our environment (Vandenberg et al. 2009). An increasing amount of scientific evidence has demonstrated diverse adverse effects of BPA on glucose homeostasis and energy balance, such as insulin resistance (Alonso-Magdalena et al. 2006; Ropero et al. 2008a), changes in body weight (Rubin et al. 2001), altered glucose transport (Sakurai et al. 2004), and altered adipocyte differentiation (Masuno et al. 2002, 2005).

In addition, we should keep in mind the hypothesis commonly named "developmental or fetal origins of adult disease," which states that adverse influences early in development and particularly during intrauterine life predict subsequent physiological disturbances in adulthood, resulting in an increased risk of developing chronic disease in adult life (Barker 1998). This paradigm is supported by extensive human epidemiological and animal model data that provide a convincing link between a non-optimal gestational environment and an increased propensity to develop adult-onset metabolic disease. It has been demonstrated that low birth weight or an accelerated postnatal weight gain or a combination of both may predispose to the risk of cardiovascular disease, hypertension and type 2 diabetes (Gilbert and Epel 2009). The most widely recognized mechanisms thought to underlie these relationships include altered fetal nutrition and increased glucocorticoid exposure (Warner and Ozanne 2010); however, the exposure to environmental hazards can also account for this complex phenomenon. It is well known that the developing fetus is extremely sensitive and can be easily disturbed by EDCs (Bern 1992; Colborn et al. 1996; Newbold et al. 2009). It has been proposed that BPA exposure during early development may be associated with increased incidence of breast cancer (Munoz-de-Toro et al. 2005), genital tract abnormalities (Skakkebaek et al. 1998) and fertility problems (Sharpe and Skakkebaek 1993). In terms of metabolism, perinatal treatment with BPA has been hypothesized to predispose to the development of diabetes (Alonso-Magdalena et al. 2010).

In this review, we will analyze the most relevant experimental and epidemiological findings that point to BPA as an important factor in the exacerbation and acceleration of metabolic disorders in the fetal and adult states.

BPA from Bench to Human Exposure

BPA was first synthesized by Dianin in 1891 and reported to be a synthetic estrogen in the 1930s (Dodds and Lawson 1936). By that time, diethylstilbestrol (DES) was also tested because of its estrogenic properties and, due to its strong estrogenic activity, BPA essentially took a back seat. In the 1950s, BPA was rediscovered as a compound that could be polymerized to make polycarbonate plastic, and from that moment on until now, it has been commonly used in the plastic industry. BPA is one of the highest volume chemicals produced worldwide, with over six billion pounds produced each year and over 100 tons released into the atmosphere by the yearly production (Vandenberg et al. 2009). It is used as the base compound in the manufacture of polycarbonate plastic and the resin lining of food and beverage cans, and as an additive in other widely used plastics such as polynil chloride and polyethylene terephthalate. It is present not only in food and beverage containers but also in some dental material (Olea et al. 1996). Numerous studies have found that BPA can leach from polycarbonate containers; heat and either acidic or basic conditions accelerate the hydrolysis of the ester bond linking BPA monomers, leading to a release of BPA with the concomitant potential human exposure (Kang et al. 2006; Richter et al. 2007). Indeed, the potential for BPA exposure has already been demonstrated since BPA was detected in 95% of the urine samples in USA (Calafat et al. 2005). Its concentration in human serum ranges from 0.2 to 1.6 ng/mL (0.88–7.0 nM; Sajiki et al. 1999; Takeuchi and Tsutsumi 2002). Moreover, it has been detected in amniotic fluid, neonatal blood, placenta, cord blood and human breast milk (Richter et al. 2007).

Concerning the potential risk of this compound, in the 1980s the lowest-observable-adverse effect-level (LOAEL) for BPA was determined at 50 mg/kg bw/day, and the Environmental Protection Agency (EPA) calculated a "reference dose" or safe dose of 50 μg/kg bw/day. However, since that time, considerable scientific evidence supports that BPA can interfere with the endocrine signaling pathways at doses below the calculated safe dose, particularly after fetal, neonatal or perinatal exposure, but also after adult exposure. A review by Richter et al. (2007) provides a comprehensive account of the findings from in vivo studies of BPA exposure.

The Concept of Metabolic Syndrome

The term "metabolic syndrome" is frequently used in research and clinical practice to describe a profile of individuals at very high risk of cardiovascular disease and diabetes mellitus. It was proposed for the first time by Reaven in a 1988 Banting lecture (Kahn et al. 2005; Reaven 1988). Reaven postulated that insulin resistance and the compensatory hyperinsulinemia were the main underlying factors in the development of diabetes, hypertension and coronary artery disease. Over time, a pathophysiological condition named "metabolic" or "insulin resistance" syndrome would be proposed as a cluster of several abnormalities such as glucose intolerance, hypertension, obesity and dyslipidemia, in which insulin resistance is the central feature. The metabolic syndrome is increasing in prevalence at alarming rates. It is estimated that around 20–25% of the world's adult population is affected. In addition, people with this condition have a fivefold greater risk of developing type 2 diabetes. There are no unified criteria for diagnosing this syndrome in children and adolescents; however, numerous studies suggest that a considerable percentage of overweight children and adolescents have one or more of the following symptoms: high triglyceride level, low HDL cholesterol level and high blood pressure. Moreover, many of them suffer hyperinsulinemia, indicating an increase in insulin resistance (Falkner et al. 2002; Freedman et al. 1999). Thus, it is estimated that nearly one million adolescents in the US (about 4%) have signs and symptoms of metabolic syndrome. Among overweight adolescents, the prevalence is nearly 30% (Cook et al. 2003).

The underlying cause of this pathology continues to be not well understood, but both insulin resistance and central obesity seem to play important roles (Fritsche et al. 2008; Herman and Kahn 2006; Ropero et al. 2008b; Rosen and Spiegelman 2006).

BPA and Metabolic Syndrome

Adult Exposure to BPA: Implications for Obesity and Insulin Resistance

A persistent imbalance in the body's energy equation – caloric intake versus expenditure – is the main mechanism that promotes obesity. It has become one of the greatest concerns in public health because it increases the risks and prognosis for a number of serious medical conditions. On an epidemiological level, there is a clear association of obesity with type 2 diabetes and the principal basis for this link is the ability of obesity to engender insulin resistance (Kahn et al. 2006). In turn, insulin resistance is thought to be the hallmark of the metabolic syndrome since this is the "driving force" of metabolic diseases. It is regarded as a state or condition characterized by diminished insulin sensitivity, meaning that the efficiency of insulin to promote glucose uptake in muscle and adipose tissue or to inhibit glucose production from the liver is decreased. Most frequently, this situation is solved by an increase in pancreatic insulin release, thereby maintaining normal glucose tolerance. However, if this compensation fails, hyperglycemia appears, which leads in turn to the development of type 2 diabetes (Biddinger and Kahn 2006;

Kahn 2003; Kahn et al. 2006, 2009). The etiology of this global problem remains unknown and most of the interest has been focused on behavioral patterns, such as excessive caloric diet and inadequate physical activity, which coexist with a genetic predisposition. Interestingly, one environmental factor that has recently received attention is the contribution of EDCs to the high prevalence of this phenomenon (Heindel and vom Saal 2009; vom Saal and Myers 2008). In this context, BPA has been shown to have a negative impact on energy balance and glucose homeostasis in adulthood (Nadal et al. 2009; Ropero et al. 2008a).

Several ex vivo studies have established a relation between BPA exposure and an altered adipocyte differentiation. Thus, it has been shown that BPA at a concentration of 10–80 μmol/L, stimulates adipogenesis in 3T3-L1 adipocytes (Masuno et al. 2002, 2005). Sargis et al. (2010) have demonstrated that 100 nM BPA is able to increase lipid accumulation in the differentiating adipocytes and upregulate the expression of adipocytic proteins through the activation of the glucocorticoid receptor. Moreover, it has been shown that BPA can affect glucose transport in adipocytes. It provokes an increase in the basal and insulin-stimulated glucose transport due to an increased amount of GLUT4 (Sakurai et al. 2004). Interestingly, BPA at 1 and 10 nM concentrations inhibits adiponectin release, an important adipokine that protects humans from the metabolic syndrome and increases the release of IL-6 and TNFα (Ben-Jonathan et al. 2009; Hugo et al. 2008).

In vivo effects of BPA on energy balance yield somewhat unclear conclusions, most likely due to the wide range of administered doses. The administration of BPA via implanted mini-osmotic pumps to ovariectomized rats at a dose of 4 or 5 mg/day for 15 days has been reported to decrease body weight gain (Nunez et al. 2001); meanwhile, treatment with lower doses applied with soy-free pelleted food (8.9 or 88 μg BPA/day for 3 months) showed no effect (Seidlova-Wuttke et al. 2005).

In addition, recent reports have demonstrated that BPA can affect insulin sensitivity and alter the function of the islet of Langerhans, the physiological unit of the endocrine pancreas. The islet of Langerhans contains four different types of cells responsible for secreting insulin (β-cells), glucagon (α-cells), somatostatin (δ-cells) and pancreatic polypeptide (PP-cells). It has been shown that BPA can have a direct effect on at least two of these cell types, α-cells and β-cells. Ex vivo studies demonstrated that BPA at a concentration of 1 nM suppresses low glucose-induced intracellular calcium oscillations on α-cells. This action is characterized by a rapid onset and involves a pertussis toxin-sensitive G-protein, nitric oxide synthase, guanylate cyclase and PKG. The hormone, glucagon, is secreted in response to low glucose concentrations, enhancing the synthesis and mobilization of glucose in the liver. Moreover, it has many extra hepatic effects such as the increase of lipolysis in adipose tissue and a role in the control of satiety in the central nervous system, among others. Thus the described effect on α-cells suggests that BPA may alter both glucose and lipid metabolism (Alonso-Magdalena et al. 2005). As regards β-cells, BPA has the opposite effect; it enhances the frequency of glucose-induced intracellular calcium oscillations at the same low concentration as previously reported, 1 nM (Nadal et al. 2000). Moreover, BPA is able to increase the activation of the transcription factor CREB in a rapid manner (Quesada et al. 2002). This effect may be of great importance for β-cell physiology, since CREB activation induces insulin gene expression (Oetjen et al. 1994) and is implicated in β-cell survival (Jhala et al. 2003).

These results have been confirmed in in vivo assays and may have important physiological implications. Experiments performed in male OF-1 mice showed that a single subcutaneous injection of 10 μg/kg of BPA produced a rapid decrease of blood glucose that occurred in parallel with an increase in plasma insulin levels (Alonso-Magdalena et al. 2006). Prolonged exposures to this compound also have effects on glucose homeostasis. Accordingly, it has been reported that male mice subcutaneously injected with BPA at a dose of 100 μg/kg for a period of 4 days experienced an increase in insulin content in pancreatic β-cells compared to vehicle-treated mice. It seems that treatment with BPA did not have an effect on β-cell survival or β-cell mass, but it affected insulin gene transcription, provoking an up-regulation of the gene in an extracellularly initiated ERα-dependent manner (Alonso-Magdalena et al. 2008). At plasma level, the 4 day-treatment with BPA at a dose of 100 μg/kg/day administered by subcutaneous injection generated a post-prandial hyperinsulinemia. Paralleling this hyperinsulinemia, an impairment glucose tolerance showed up. When a glucose tolerance test was performed in BPA-treated mice, impaired glucose tolerance was observed, indicating that they were insulin resistant. This finding was confirmed by an insulin tolerance test that showed a less hypoglycemic response to a challenge of insulin in those animals treated with BPA (Alonso-Magdalena et al. 2006).

Metabolic Effects of BPA During Development

Developing organisms are extremely sensitive to BPA exposure since the liver has limited capacity to deactivate this compound in fetuses and newborns (Coughtrie et al. 1988; Takahashi and Oishi 2000). Besides, BPA elimination by hepatic glucuronidation is slightly lower in pregnancy than in non-pregnancy (Inoue et al. 2005). In addition, BPA is considered to have increased access to tissues during development due to the lack of binding to alpha fetoprotein, the major plasma estrogen binding protein that limits estradiol bioavailability (Milligan et al. 1998). Evaluating the extent to which perinatal exposure to EDCs, in particular BPA, modulates metabolic features will significantly improve our understanding of the adult origin of metabolic diseases.

The first evidence that early exposure to BPA could interfere with lipid metabolism and energy balance was published by Howdeshell et al. in Howdeshell et al. (1999). The authors demonstrated that, when pregnant CF-1 mice were fed BPA at a dose of 2.4 μg/kg on days 11–17 of gestation, their pups were heavier compared with controls on postnatal day 22. The effect was dependent on the position of the fetus, such that those mouse fetuses positioned between two female fetuses showed a greater effect. Later, it was shown that offspring from rats exposed to approximate doses of BPA of 0.1 or 1.2 mg/kg bw/day in the drinking water (from day 6 of pregnancy through the lactation period) showed an increase in body weight that was evident soon after birth and continued into adulthood (Rubin et al. 2001). In another study, pregnant mice were exposed to BPA in drinking water at 1 or 10 μg/mL from day 10 of gestation throughout the lactation period. Body and adipose tissue weights were measured 31 days after birth, showing an increase in both parameters in a sex and dose-dependent manner (Miyawaki et al. 2007). When pregnant rats received 1 mg/L BPA in drinking water from day 6 of gestation through the end of lactation, alterations in adipogenesis were observed. On postnatal day 21, female pups showed an increase in white adipose tissue that was associated with adipocyte hypertrophy and an overexpression of some adipogenic genes, including PPAR-γ, SREBP-1C, SCD-1 and C/EBP-α (Somm et al. 2009). Administration of BPA by subcutaneous injection for shorter periods at doses of 0.5 or 10 mg/kg bw/day to pregnant CD-1 mice from day 15 of gestation also resulted in an accelerated body weight gain in female offspring (Nikaido et al. 2004). With regards to postnatal BPA exposure, male neonates, just after birth, subcutaneously injected with BPA at 50 μg/kg bw/day for 4 days were heavier on postnatal day 68 (Patisaul and Bateman 2008).

Up to now, no studies have explored the effects of BPA on pregnancy outcome and focused on the potential risk of developing diabetes in mothers after labor; even so, this disease is a major potential complication of pregnancy, with adverse effects for both mothers and newborns. Alonso-Magdalena et al. (2010) studied the action of environmentally relevant doses of BPA during gestation and the consequences of in utero exposure on glucose homeostasis. The authors treated OF-1 pregnant mice with two different doses of BPA, 10 and 100 μg/kg bw/day, by subcutaneous injection on days 9–16 of gestation. They observed that the treatment with BPA aggravated the insulin resistance that is characteristic of pregnancy, producing an impairment of glucose tolerance and higher plasma insulin, triglyceride and leptin levels compared to control pregnant mice. Insulin-stimulated Akt phosphorylation was reduced in skeletal muscle and liver of BPA-treated pregnant mice. Moreover, it was evident that BPA exposure had long-term consequences for the mothers and adult male offspring. Four months after delivery, mice weighed more and showed higher plasma insulin, leptin, triglyceride and glycerol levels than untreated mice. They also showed a mild insulin resistance. With regards to the offspring, at 6 months of age male mice had reduced glucose tolerance, increased insulin resistance and altered metabolite levels, showing for the first time that low doses of BPA during critical periods of life had adverse effects on glucose homeostasis and insulin sensitivity (Alonso-Magdalena et al. 2010). In another study by Ryan et al. (2010), the susceptibility in terms of glucose sensitivity to a high fat diet was explored after perinatal exposure to BPA. Pregnant mice were fed a special diet that contained BPA, at an approximate dose of 0.25 μg/kg bw/day. They were exposed to this diet until their pups were weaned. The authors found that pups whose mothers had been exposed to BPA were heavier at weaning than controls. Moreover, at 4 weeks of age BPA mice were longer than control mice, although these differences disappeared when they reached adulthood. No effects on glucose tolerance were observed at 8 weeks of age. At the age of 9 weeks, the diet was changed to a high fat diet to explore whether perinatal exposure to BPA might increase the susceptibility to diet-induced obesity; however, no differences in body weight or glucose tolerance were observed in BPA mice when compared with controls (Ryan et al. 2010).

Possible Mechanisms of Action of BPA

Traditionally BPA has been considered to be a weak environmental estrogen since its binding affinity to estrogen receptor alfa (ERα) and estrogen receptor beta (ERβ; Kuiper et al. 1998) is estimated to be over 1,000–10,000-fold lower than the natural hormone 17β-estradiol (E2; Kuiper et al. 1998). However, recent studies have revealed that BPA can promote estrogen-like activities that are similar or stronger than E2 (Alonso-Magdalena et al. 2005, 2006; Hugo et al. 2008; Zsarnovszky et al. 2005). These low dose effects can be explained at least partially because BPA elicits rapid responses via non-classical estrogen triggered pathways (Nadal et al. 2000; Quesada et al. 2002; Watson et al. 2005). In addition, it may bind differently than E2 within the ligand domain of estrogen receptors (ERs; Gould et al. 1998) as well as recruit different co-regulators (Routledge et al. 2000; Safe et al. 2002).

As previously described, BPA has been shown to alter pancreatic function in vitro and in vivo. Within pancreatic β−cells, low doses of BPA increase insulin content and insulin secretion. This effect has been found to be mediated via extranuclear activation of ERα and extracellular signal-regulated kinases (1/2) (Alonso-Magdalena et al. 2008).

In addition, BPA has been demonstrated to bind to the estrogen related receptor ERRγ (Matsushima et al. 2008; Okada et al. 2008). ERRγ is present in adipose tissue, where it possibly plays an important role in the control of energy homeostasis (Giguere 2008). It remains unknown if BPA may exert effects on body weight regulation via this receptor.

Apart from its estrogenic activities, BPA can bind to the thyroid hormone receptor and inhibit thyroid hormone receptor-mediated gene expression by enhancing the recruitment of the co-repressor N-CoR to the thyroid hormone receptor (Moriyama et al. 2002; Xu et al. 2007; Zoeller et al. 2005). This may have subsequent implications because of the role that thyroid hormones plays in energy homeostasis.

Epidemiological Evidence

Most of the evidence that relates obesity and type 2 diabetes with the exposure to BPA comes from laboratory studies; thus it is a matter of debate whether the outcomes can be extrapolated to humans. We here review the epidemiological data available.

A recent cross-sectional analysis of spot urinary BPA concentrations and health status in a representative sample of the adult population of the United States in 2003–2004 (the National Health and Nutrition Examination Survey (NHANES)) showed that higher urinary levels of BPA were associated with abnormal levels of the liver enzymes γ-glutamyl-transferase, alkaline phosphatase and lactate dehydrogenase. The authors also established a significant relationship between BPA concentration and type 2 diabetes and cardiovascular disease (Lang et al. 2008).

A similar study was carried out using later NHANES data from 2005 to 2006. The authors found that urinary BPA concentrations were significantly lower in 2005–2006 (1.79 ng/mL) than in the population of 2003–2004 (2.49 ng/mL). However, a correlation between urinary BPA concentrations and an increased prevalence of coronary heart disease was still present. With regards to the prevalence of diabetes and liver enzyme abnormalities, the association with urinary BPA was not significant but the pooled estimate remained significant (Melzer et al. 2010).

Conclusions

The data reviewed above reveal that EDCs, in particular BPA, alter glucose and lipid metabolism in animal models. A brief exposure to low doses of BPA can alter insulin sensitivity and glucose tolerance in adult mice as well as decrease adiponectin or increase IL-6 and TNF-α in human adipocytes. The administration of this compound early in development promotes a pre-diabetic state in adult male offspring. In human adults, epidemiological evidence supports the findings in animals and points to BPA as an important risk factor for type 2 diabetes. Therefore, BPA exposure should be regarded as an exacerbating and accelerating factor in the development of metabolic disorders in human beings.

References

Alonso-Magdalena P, Laribi O, Ropero AB, Fuentes E, Ripoll C, Soria B, Nadal A (2005) Low doses of bisphenol A and diethylstilbestrol impair Ca^{2+} signals in pancreatic alpha-cells through a nonclassical membrane estrogen receptor within intact islets of Langerhans. Environ Health Perspect 113:969–977

Alonso-Magdalena P, Morimoto S, Ripoll C, Fuentes E, Nadal A (2006) The estrogenic effect of bisphenol A disrupts pancreatic beta-cell function in vivo and induces insulin resistance. Environ Health Perspect 114:106–112

Alonso-Magdalena P, Ropero AB, Carrera MP, Cederroth CR, Baquie M, Gauthier BR, Nef S, Stefani E, Nadal A (2008) Pancreatic insulin content regulation by the estrogen receptor ER alpha. PLoS One 3(4):e2069

Alonso-Magdalena P, Vieira E, Soriano S, Menes L, Burks D, Quesada I, Nadal A (2010) Bisphenol-A exposure during pregnancy disrupts glucose homeostasis in mothers and adult male offspring. Environ Health Perspect 118:1243–1250

Barker DJ (1998) In utero programming of chronic disease. Clin Sci (London) 95:115–128

Ben-Jonathan N, Hugo ER, Brandebourg TD (2009) Effects of bisphenol A on adipokine release from human adipose tissue: implications for the metabolic syndrome. Mol Cell Endocrinol 304:49–54

Bern B (1992) The fragil fetus. Princeton Scientific Publishing Co., Princenton, NJ

Biddinger SB, Kahn CR (2006) From mice to men: insights into the insulin resistance syndromes. Annu Rev Physiol 68:123–158

Calafat AM, Kuklenyik Z, Reidy JA, Caudill SP, Ekong J, Needham LL (2005) Urinary concentrations of bisphenol A and 4-nonylphenol in a human reference population. Environ Health Perspect 113:391–395

Colborn T, Dumanoski D, Myers JP (1996) Our stolen future: are we threatening our fertility, intelligence and survival. Penguin Books, New York

Cook S, Weitzman M, Auinger P, Nguyen M, Dietz WH (2003) Prevalence of a metabolic syndrome phenotype in adolescents: findings from the third National Health and Nutrition Examination Survey, 1988–1994. Arch Pediatr Adolesc Med 157:821–827

Coughtrie MW, Burchell B, Leakey JE, Hume R (1988) The inadequacy of perinatal glucuronidation: immunoblot analysis of the developmental expression of individual UDP-glucuronosyltransferase isoenzymes in rat and human liver microsomes. Mol Pharmacol 34(6):729–735

Dabelea D, Mayer-Davis EJ, Imperatore G (2010) The value of national diabetes registries: SEARCH for diabetes in youth study. Curr Diab Rep 10:362–369

Dodds EC, Lawson W (1936) Synthetic estrogenic agents without the phenanthrene nucleus. Nature 137:996

Falkner B, Hassink S, Ross J, Gidding S (2002) Dysmetabolic syndrome: multiple risk factors for premature adult disease in an adolescent girl. Pediatrics 110(1 Pt 1):e14

Fox CS, Pencina MJ, Meigs JB, Vasan RS, Levitzky YS, D'Agostino RB Sr (2006) Trends in the incidence of type 2 diabetes mellitus from the 1970s to the 1990s: the Framingham heart study. Circulation 113:2914–2918

Freedman DS, Dietz WH, Srinivasan SR, Berenson GS (1999) The relation of overweight to cardiovascular risk factors among children and adolescents: the Bogalusa Heart Study. Pediatrics 103(6 Pt 1):1175–1182

Fritsche L, Weigert C, Haring HU, Lehmann R (2008) How insulin receptor substrate proteins regulate the metabolic capacity of the liver–implications for health and disease. Curr Med Chem 15:1316–1329

Giguere V (2008) Transcriptional control of energy homeostasis by the estrogen-related receptors. Endocr Rev 29:677–696

Gilbert S, Epel D (2009) Ecological developmental biology: integrating epigenetics, medicine and evolution. Sinauer Associates Inc., Sunderland, MA

Gould JC, Leonard LS, Maness SC, Wagner BL, Conner K, Zacharewski T, Safe S, McDonnell DP, Gaido KW (1998) Bisphenol A interacts with the estrogen receptor alpha in a distinct manner from estradiol. Mol Cell Endocrinol 142:203–214

Grandjean P, Landrigan PJ (2006) Developmental neurotoxicity of industrial chemicals. Lancet 368:2167–2178

Heindel JJ, Vom Saal FS (2009) Role of nutrition and environmental endocrine disrupting chemicals during the perinatal period on the aetiology of obesity. Mol Cell Endocrinol 304:90–96

Heintz NH, Janssen-Heininger YM, Mossman BT (2010) Asbestos, lung cancers, and mesotheliomas: from molecular approaches to targeting tumor survival pathways. Am J Respir Cell Mol Biol 42:133–139

Herman MA, Kahn BB (2006) Glucose transport and sensing in the maintenance of glucose homeostasis and metabolic harmony. J Clin Invest 116:1767–1775

Howdeshell KL, Hotchkiss AK, Thayer KA, Vandenbergh JG, Vom Saal FS (1999) Exposure to bisphenol A advances puberty. Nature 401:763–764

Hugo ER, Brandebourg TD, Woo JG, Loftus J, Alexander JW, Ben-Jonathan N (2008) Bisphenol A at environmentally relevant doses inhibits adiponectin release from human adipose tissue explants and adipocytes. Environ Health Perspect 116:1642–1647

Inoue H, Tsuruta A, Kudo S, Ishii T, Fukushima Y, Iwano H, Yokota H, Kato S (2005) Bisphenol a glucuronidation and excretion in liver of pregnant and nonpregnant female rats. Drug Metab Dispos 33:55–59

Ito T, Ando H, Suzuki T, Ogura T, Hotta K, Imamura Y, Yamaguchi Y, Handa H (2010) Identification of a primary target of thalidomide teratogenicity. Science 327:1345–1350

Jhala US, Canettieri G, Screaton RA, Kulkarni RN, Krajewski S, Reed J, Walker J, Lin X, White M, Montminy M (2003) cAMP promotes pancreatic beta-cell survival via CREB-mediated induction of IRS2. Genes Dev 17:1575–1580

Kahn CR (2003) Knockout mice challenge our concepts of glucose homeostasis and the pathogenesis of diabetes. Exp Diabesity Res 4:169–182

Kahn R, Buse J, Ferrannini E, Stern M (2005) The metabolic syndrome: time for a critical appraisal: joint statement from the American Diabetes Association and the European Association for the Study of Diabetes. Diabetes Care 28:2289–2304

Kahn SE, Hull RL, Utzschneider KM (2006) Mechanisms linking obesity to insulin resistance and type 2 diabetes. Nature 444:840–846

Kahn SE, Zraika S, Utzschneider KM, Hull RL (2009) The beta cell lesion in type 2 diabetes: there has to be a primary functional abnormality. Diabetologia 52:1003–1012

Kang JH, Kondo F, Katayama Y (2006) Human exposure to bisphenol A. Toxicology 226:79–89

Kuiper GG, Lemmen JG, Carlsson B, Corton JC, Safe SH, van der Saag PT, van der Burg B, Gustafsson JA (1998) Interaction of estrogenic chemicals and phytoestrogens with estrogen receptor beta. Endocrinology 139:4252–4263

Lang IA, Galloway TS, Scarlett A, Henley WE, Depledge M, Wallace RB, Melzer D (2008) Association of urinary bisphenol A concentration with medical disorders and laboratory abnormalities in adults. JAMA 300:1303–1310

Masuno H, Kidani T, Sekiya K, Sakayama K, Shiosaka T, Yamamoto H, Honda K (2002) Bisphenol A in combination with insulin can accelerate the conversion of 3T3-L1 fibroblasts to adipocytes. J Lipid Res 43:676–684

Masuno H, Iwanami J, Kidani T, Sakayama K, Honda K (2005) Bisphenol a accelerates terminal differentiation of 3T3-L1 cells into adipocytes through the phosphatidylinositol 3-kinase pathway. Toxicol Sci 84:319–327

Matsushima A, Teramoto T, Okada H, Liu X, Tokunaga T, Kakuta Y, Shimohigashi Y (2008) ERRgamma tethers strongly bisphenol A and 4-alpha-cumylphenol in an induced-fit manner. Biochem Biophys Res Commun 373:408–413

Melzer D, Rice NE, Lewis C, Henley WE, Galloway TS (2010) Association of urinary bisphenol a concentration with heart disease: evidence from NHANES 2003/06. PLoS One 5(1):e8673

Milligan SR, Khan O, Nash M (1998) Competitive binding of xenobiotic oestrogens to rat alpha-fetoprotein and to sex steroid binding proteins in human and rainbow trout (*Oncorhynchus mykiss*) plasma. Gen Comp Endocrinol 112:89–95

Miyawaki J, Sakayama K, Kato H, Yamamoto H, Masuno H (2007) Perinatal and postnatal exposure to bisphenol a increases adipose tissue mass and serum cholesterol level in mice. J Atheroscler Thromb 14:245–252

Moriyama K, Tagami T, Akamizu T, Usui T, Saijo M, Kanamoto N, Hataya Y, Shimatsu A, Kuzuya H, Nakao K (2002) Thyroid hormone action is disrupted by bisphenol A as an antagonist. J Clin Endocrinol Metab 87:5185–5190

Munoz-de-Toro M, Markey CM, Wadia PR, Luque EH, Rubin BS, Sonnenschein C, Soto AM (2005) Perinatal exposure to bisphenol-A alters peripubertal mammary gland development in mice. Endocrinology 146:4138–4147

Nadal A, Ropero AB, Laribi O, Maillet M, Fuentes E, Soria B (2000) Nongenomic actions of estrogens and xenoestrogens by binding at a plasma membrane receptor unrelated to estrogen receptor alpha and estrogen receptor beta. Proc Natl Acad Sci USA 97:11603–11608

Nadal A, Alonso-Magdalena P, Soriano S, Quesada I, Ropero AB (2009) The pancreatic beta-cell as a target of estrogens and xenoestrogens: implications for blood glucose homeostasis and diabetes. Mol Cell Endocrinol 304:63–68

Newbold RR, Padilla-Banks E, Jefferson WN, Heindel JJ (2008) Effects of endocrine disruptors on obesity. Int J Androl 31:201–208

Newbold RR, Padilla-Banks E, Jefferson WN (2009) Environmental estrogens and obesity. Mol Cell Endocrinol 304:84–89

Nikaido Y, Yoshizawa K, Danbara N, Tsujita-Kyutoku M, Yuri T, Uehara N, Tsubura A (2004) Effects of maternal xenoestrogen exposure on development of the reproductive tract and mammary gland in female CD-1 mouse offspring. Reprod Toxicol 18:803–811

Nunez AA, Kannan K, Giesy JP, Fang J, Clemens LG (2001) Effects of bisphenol A on energy balance and accumulation in brown adipose tissue in rats. Chemosphere 42:917–922

O'Rahilly S (2009) Human genetics illuminates the paths to metabolic disease. Nature 462:307–314

Oetjen E, Diedrich T, Eggers A, Eckert B, Knepel W (1994) Distinct properties of the cAMP-responsive element of the rat insulin I gene. J Biol Chem 269:27036–27044

Okada H, Tokunaga T, Liu X, Takayanagi S, Matsushima A, Shimohigashi Y (2008) Direct evidence revealing structural elements essential for the high binding ability of bisphenol A to human estrogen-related receptor-gamma. Environ Health Perspect 116:32–38

Olea N, Pulgar R, Perez P, Olea-Serrano F, Rivas A, Novillo-Fertrell A, Pedraza V, Soto AM, Sonnenschein C (1996) Estrogenicity of resin-based composites and sealants used in dentistry. Environ Health Perspect 104:298–305

Patisaul HB, Bateman HL (2008) Neonatal exposure to endocrine active compounds or an ERbeta agonist increases adult anxiety and aggression in gonadally intact male rats. Horm Behav 53:580–588

Qatanani M, Lazar MA (2007) Mechanisms of obesity-associated insulin resistance: many choices on the menu. Genes Dev 21:1443–1455

Quesada I, Fuentes E, Viso-Leon MC, Soria B, Ripoll C, Nadal A (2002) Low doses of the endocrine disruptor bisphenol-A and the native hormone 17beta-estradiol rapidly activate transcription factor CREB. FASEB J 16:1671–1673

Reaven GM (1988) Banting lecture 1988. Role of insulin resistance in human disease. Diabetes 37:1595–1607

Richter CA, Birnbaum LS, Farabollini F, Newbold RR, Rubin BS, Talsness CE, Vandenbergh JG, Walser-Kuntz DR, Vom Saal FS (2007) In vivo effects of bisphenol A in laboratory rodent studies. Reprod Toxicol 24:199–224

Ropero AB, Alonso-Magdalena P, Garcia-Garcia E, Ripoll C, Fuentes E, Nadal A (2008a) Bisphenol-A disruption of the endocrine pancreas and blood glucose homeostasis. Int J Androl 31:194–200

Ropero AB, Alonso-Magdalena P, Quesada I, Nadal A (2008b) The role of estrogen receptors in the control of energy and glucose homeostasis. Steroids 73:874–879

Rosen ED, Spiegelman BM (2006) Adipocytes as regulators of energy balance and glucose homeostasis. Nature 444(7121):847–853

Rosenbloom AL, Joe JR, Young RS, Winter WE (1999) Emerging epidemic of type 2 diabetes in youth. Diabetes Care 22:345–354

Routledge EJ, White R, Parker MG, Sumpter JP (2000) Differential effects of xenoestrogens on coactivator recruitment by estrogen receptor (ER) alpha and ERbeta. J Biol Chem 275:35986–35993

Rubin BS, Murray MK, Damassa DA, King JC, Soto AM (2001) Perinatal exposure to low doses of bisphenol A affects body weight, patterns of estrous cyclicity, and plasma LH levels. Environ Health Perspect 109:675–680

Ryan KK, Haller AM, Sorrell JE, Woods SC, Jandacek RJ, Seeley RJ (2010) Perinatal exposure to bisphenol-a and the development of metabolic syndrome in CD-1 mice. Endocrinology 151:2603–2612

Safe SH, Pallaroni L, Yoon K, Gaido K, Ross S, McDonnell D (2002) Problems for risk assessment of endocrine-active estrogenic compounds. Environ Health Perspect 110(Suppl 6):925–929

Sajiki J, Takahashi K, Yonekubo J (1999) Sensitive method for the determination of bisphenol-A in serum using two systems of high-performance liquid chromatography. J Chromatogr B Biomed Sci Appl 736:255–261

Sakurai K, Kawazuma M, Adachi T, Harigaya T, Saito Y, Hashimoto N, Mori C (2004) Bisphenol A affects glucose transport in mouse 3T3-F442A adipocytes. Br J Pharmacol 141:209–214

Sargis RM, Johnson DN, Choudhury RA, Brady MJ (2010) Environmental endocrine disruptors promote adipogenesis in the 3T3-L1 cell line through glucocorticoid receptor activation. Obesity (Silver Spring) 18:1283–1288

Seidlova-Wuttke D, Jarry H, Christoffel J, Rimoldi G, Wuttke W (2005) Effects of bisphenol-A (BPA), dibutylphtalate (DBP), benzophenone-2 (BP2), procymidone (Proc), and linurone (Lin) on fat tissue, a variety of hormones and metabolic parameters: a 3 months comparison with effects of estradiol (E2) in ovariectomized (ovx) rats. Toxicology 213:13–24

Sharpe RM, Skakkebaek NE (1993) Are oestrogens involved in falling sperm counts and disorders of the male reproductive tract? Lancet 341:1392–1395

Skakkebaek NE, Rajpert-De Meyts E, Jorgensen N, Carlsen E, Petersen PM, Giwercman A, Andersen AG, Jensen TK, Andersson AM, Müller J (1998) Germ cell cancer and disorders of spermatogenesis: an environmental connection? Apmis 106:3–11, discussion 12

Somm E, Schwitzgebel VM, Toulotte A, Cederroth CR, Combescure C, Nef S, Aubert ML, Hüppi PS (2009) Perinatal exposure to bisphenol a alters early adipogenesis in the rat. Environ Health Perspect 117:1549–1555

Takahashi O, Oishi S (2000) Disposition of orally administered 2,2-Bis(4-hydroxyphenyl)propane (bisphenol A) in pregnant rats and the placental transfer to fetuses. Environ Health Perspect 108:931–935

Takeuchi T, Tsutsumi O (2002) Serum bisphenol a concentrations showed gender differences, possibly linked to androgen levels. Biochem Biophys Res Commun 291:76–78

Vandenberg LN, Maffini MV, Sonnenschein C, Rubin BS, Soto AM (2009) Bisphenol-A and the great divide: a review of controversies in the field of endocrine disruption. Endocr Rev 30:75–95

Vom Saal FS, Myers JP (2008) Bisphenol A and risk of metabolic disorders. JAMA 300:1353–1355

Warner MJ, Ozanne SE (2010) Mechanisms involved in the developmental programming of adulthood disease. Biochem J 427:333–347

Watson CS, Bulayeva NN, Wozniak AL, Finnerty CC (2005) Signaling from the membrane via membrane estrogen receptor-alpha: estrogens, xenoestrogens, and phytoestrogens. Steroids 70:364–371

Xu X, Liu Y, Sadamatsu M, Tsutsumi S, Akaike M, Ushijima H, Kato N (2007) Perinatal bisphenol A affects the behavior and SRC-1 expression of male pups but does not influence on the thyroid hormone receptors and its responsive gene. Neurosci Res 58:149–155

Zoeller RT, Bansal R, Parris C (2005) Bisphenol-A, an environmental contaminant that acts as a thyroid hormone receptor antagonist in vitro, increases serum thyroxine, and alters RC3/neurogranin expression in the developing rat brain. Endocrinology 146:607–612

Zsarnovszky A, Le HH, Wang HS, Belcher SM (2005) Ontogeny of rapid estrogen-mediated extracellular signal-regulated kinase signaling in the rat cerebellar cortex: potent nongenomic agonist and endocrine disrupting activity of the xenoestrogen bisphenol A. Endocrinology 146:5388–5396

Bisphenol A in the Gut: Another Break in the Wall?

Viorica Braniste, Marc Audebert, Daniel Zalko, and Eric Houdeau

Abstract From animal studies, a consensus exists that the synthetic estrogen bisphenol A (BPA), a plastic monomer widely used in the food-packaging industry, is able to disrupt endocrine signalling pathways during development, with persisting effects later in life. Although the fetal and then the adult gut expresses functional estrogen receptors (ERs), the endocrine impact of BPA on the intestinal barrier function remains largely unexplored. The intestinal epithelium and mucosal immune cells provide a first line of defence designed to restrict the passage of harmful substances from the lumen. Intestinal permeability is high at birth, permitting lumen-to-mucosa exchanges involved in the maturation process of the gut immune system. As a barrier to the external environment, gut epithelium is renewed constantly during life. Renewal of gut epithelial cells occurs in less than ~96 h, starting from fetal stage, and is dependent on controlled cell stimulation and proliferation by various signalling pathways. Lessons learned from ER-deficient mice underline the importance of estrogen signalling in growth, organization and maintenance of a normal epithelial barrier. In rats, BPA was recently shown to interact with ERs in the adult gut by mimicking the estradiol-mediated decrease of epithelial permeability through genomic pathways. This effect also occurs in neonates when low doses of BPA (5 μg/kg BW/day: tenfold below the tolerable daily intake for humans) are orally administered to pregnant and then lactating rats. A perinatal exposure to BPA also reduces epithelial cell proliferation in the colon of neonates, while the overall decrease of intestinal permeability remains apparent in adulthood only in female offspring. As a consequence, adult females perinatally exposed to BPA have been shown to develop severe inflammatory responses in a rat model of inflammatory bowel disease, demonstrating enhanced expression and production of T-Helper 1 cytokines in inflamed areas. In mice, a mother-to-infant

E. Houdeau (✉)
Neurogastroenterology & Nutrition Unit, UMR 1054, Institut National de la Recherche Agronomique, ToxAlim, Toulouse, France
e-mail: eric.houdeau@toulouse.inra.fr

J.-P. Bourguignon et al. (eds.), *Multi-System Endocrine Disruption*, Research and Perspectives in Endocrine Interactions,
DOI 10.1007/978-3-642-22775-2_9, © Springer-Verlag Berlin Heidelberg 2011

transport of maternal BPA is consistent with the ability of the chemical to reach the fetus through the placental barrier: BPA is present in the amniotic fluid and accumulates in the maturing gut. Although BPA, once absorbed by maternal gut, is rapidly deactivated by first pass conjugation in the liver, recent studies emphasize that BPA at low, environmentally relevant levels can transfer across the human placenta, mainly in an estrogen-active, unconjugated form. It is now thought that BPA ingested by dams has repercussions on the education of the immune system by reducing intestinal permeability from fetal stages and promotes severe inflammatory response later in life.

Introduction

The intestine is the first organ in contact with ingested food. It should allow the absorption of nutrients and water, which are necessary for growth and body fluid homeostasis, but at the same time, it should protect the host against the entry of undesirable factors (Masyuk et al. 2002; Turner 2009). The intestinal epithelium is in continuous contact with dietary antigens and microorganisms, including normal microbiota and pathogenic bacteria that constantly stimulate mucosal immune cells. In normal conditions, this contact results in tolerance of the commensal flora and protection against pathogenic shock (Wagner et al. 2008). The gut barrier function develops early in life, organized around a trophic epithelium in constant renewal, a protective layer of mucus lining epithelial cells, and the mucosal immune cells, incompetent at birth, that need to develop by discriminating between pathogenic antigens and the microflora that colonize the digestive tract. A dysregulation in this process of immune acquisition and tolerance may lead to the development of chronic mucosal injury, functional gastrointestinal disorders, or tumors (Wagner et al. 2008; Turner 2009).

The digestive system is a newly identified target for estrogens (Enmark et al. 1997; Pfaffl et al. 2003; Houdeau et al. 2007; O'Mahony and Harvey 2008; Braniste et al. 2009). Although estrogen receptors (ERs) alpha and beta are expressed in epithelial and immune cells (Campbell-Thompson et al. 2001; Konstantinopoulos et al. 2003), the gut is rarely considered to be sensitive to endocrine disruptors ingested with food. Recent studies highlighted that endogenous estrogens play a role in the renewal of enterocytes (Wada-Hiraike et al. 2006), intestinal permeability (Braniste et al. 2009) or the immune response to gastrointestinal inflammation (Verdu et al. 2002; Harnish et al. 2004; Houdeau et al. 2007), findings that raised the question of the influence of xenoestrogens on the integrity of the intestinal barrier function. The pleiotropic effects of estrogens in other biological barriers of the body such as skin (Emmerson et al. 2009), lung (Speyer et al. 2005) and the blood-brain barrier (Wilson et al. 2008), as well as the variation in intestinal pain perception during the menstrual cycle (Heitkemper et al. 2003), have opened a new field of study on the influence of hormonal status in the pathophysiology of the intestine and, more recently, have led to questioning of the putative impacts of xenoestrogen contaminants present in food.

Among these contaminants, many studies have focused their attention on the effects of the plastic monomer and plasticizer bisphenol A (BPA) on various estrogen-sensitive functions, including at low doses of exposure. Despite a weak estrogenic activity (Kuiper et al. 1997, 1998; Kim et al. 2001; Matthews et al. 2001), many scientists consider the perinatal period as a critical window of exposure for BPA, considering the hypersensitivity of developing tissues to estrogens during fetal life and young infancy (Aksglaede et al. 2006; Bondesson et al. 2009). Recent data have pointed out the need for an accurate assessment of the impact of BPA on the digestive tract, with a focus on the gut barrier function and its development. Of particular interest is the maturation of the mucosal immune system, guaranteeing an effective defence for life against undesirable factors.

Human Exposure to BPA: Intake from Food and Beverages

The xenoestrogen BPA is the building monomer of polycarbonate plastics used to manufacture baby bottles, drinking containers or tableware (plates and mugs) (Brede et al. 2003; Burridge 2003). In addition, BPA-based epoxy resins serve as coatings of food and beverage metal cans, both to protect the canned food against leaching of metals or corrosion during heat treatment for microbe destruction and for long storage by consumers (Brotons et al. 1995; Howe and Borodinsky 1998). BPA is also used as an additive in other plastics, such as polyvinyl chloride (PVC) to manufacture water pipes, as well as stretch films in food packaging (Lopez-Cervantes and Paseiro-Losada 2003). Furthermore, the presence of BPA and closely related analogs (*bisphenol A diglycidyl ether* or BADGE) in some dental sealants is considered to be a non-negligible source for oral intake (Olea et al. 1996). Whatever the source, free BPA has been shown to leach from these polymers under normal conditions (Carwile et al. 2009; Cao et al. 2010; Nam et al. 2010), whereas exposure of food and beverage containers to elevated temperatures (boiling, heating) or basic or acidic conditions greatly increase its rate of migration from polycarbonated plastics and epoxy resins (Brede et al. 2003; Sajiki and Yonekubo 2004). Because of the wide use of BPA in the food-packaging industry, diet is considered the major source of exposure in humans (WHO 2010). As a protective measure, the European Food Safety Agency and US Environmental Protection Agency have established, based on classical toxicological studies, a Tolerable human Daily Intake (TDI) of 50 μg/kg of body weight (BW)/day, applying an uncertainty factor of 100 to the No Observed Adverse Effect Level (NOAEL) of 5 mg/kg BW/day in animal models. Despite these safety rules, studies have shown that BPA is recovered in 95% of human urine samples in industrialized countries (Calafat et al. 2005, 2008) and is detected in serum, umbilical cord blood, amniotic fluid, and fetal tissues in a range of concentrations predicted to be biologically active on ERs and able to interfere with normal sex hormone balance (Vandenberg et al. 2007; Bondesson et al. 2009). Although the affinity of BPA for ERs (both alpha and beta ER) is low – estimated to be 10,000–100,000 weaker than that of the mere hormone estradiol – recent data in human cell models and rodents highlighted that BPA can trigger a variety of cellular responses at very low exposure levels (Welshons et al. 2006). This is the case for the gut, where BPA below the current reference dose (TDI) appeared equally as potent as estradiol in evoking changes in intestinal barrier function, through binding of BPA to ERs (Braniste et al. 2010). The results of this study emphasized that BPA could have potent effects at low doses in the gastrointestinal tract, leading to the assumption that free BPA ingested with food may be potent in disrupting estrogen-sensitive functions in the gut, even before its absorption and distribution to other organs.

Absorption, Biotransformation and Disposition of BPA: From Mother to Infant Gut

Conjugated and Unconjugated BPA in the Organism: A Subtle Balance for Estrogenic Effects

Analysis of metabolic pathways of endocrine disrupting chemicals after dietary exposure is essential to determine their effects in the body, beginning with the contact with the intestinal barrier. The biological activity of these chemicals is modified when the parent compound is metabolized through detoxication pathways in the body. The example of BPA is particularly interesting. After absorption, a first-pass hepatic passage occurs before systemic circulation. In the liver, UDP glucuronyl transferases (UDPGT) conjugate BPA with glucuronic acid, transforming free parent BPA into BPA-glucuronide (BPA-Gluc) (Fig. 1). BPA-Gluc is biologically inactive on ERs (Matthews et al. 2001). However, and to a certain extent, BPA-Gluc can be deconjugated back into parent BPA, since many tissues including the gut express functional ß-glucuronidase activities in human as well as in animals (Borghoff and Birnbaum 1985; Sperker et al. 1997; O'Leary et al. 2003). Finally, a key issue regarding the bioactivity of BPA in the organism is a better understanding of the extent to which BPA-Gluc is reactivated through hydrolysis at the level of estrogen-sensitive tissues.

At the level of the gut, there is also evidence in rats that BPA is glucuronidated during its absorption by the intestine and contributes with the liver to the entero-hepatic cycle of detoxication of the compound before systemic circulation (Inoue et al. 2000, 2003). However, studies performed on everted intestine in rats also revealed that a non-negligible amount of BPA added on the mucosal side remained free after 60 min of incubation (Inoue et al. 2003). Additional experiments by Audebert and co-workers showed that large amounts of free BPA (50% of initial concentration) remained in contact with cultured LS-174T human colonocytes 4 h after the beginning of incubation with BPA (unpublished data), and there were very poor BPA conjugation rates within 1 h (Fig. 2). It is thus reasonable to consider that a significant amount of free BPA is present in the intestinal lumen after oral

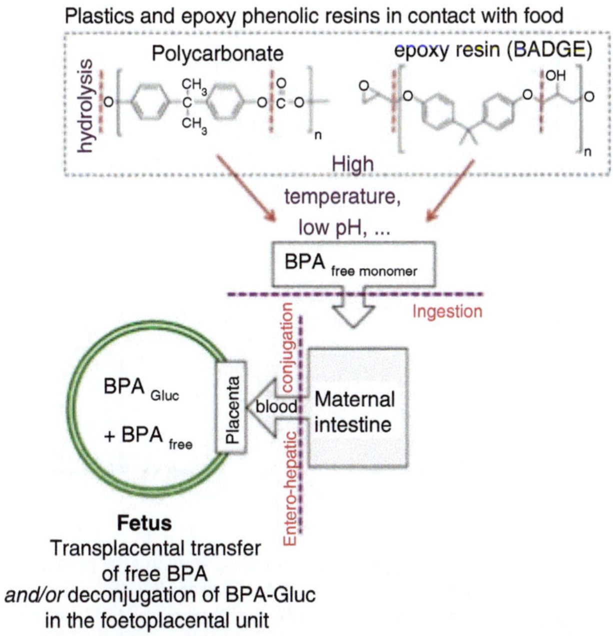

Fig. 1 Chemical structures of polycarbonate and epoxy resins, and main degradation pathways by hydrolysis of ester bond linking BPA monomers. Free molecule ingested with food is absorbed by the intestine and the liver and conjugated by a first-pass hepatic passage to form a BPA-glucuronide (BPA-Gluc) devoid of estrogenic properties. During gestation, BPA from maternal blood crosses the placental barrier and is recovered in amniotic fluid and fetal tissues. Both the placenta and fetus have the potential to cleave BPA-Gluc for local re-activation of BPA in free form (see text)

exposure and is able to stimulate local cellular responses through ERs expressed by epithelial cells.

Disposition and Distribution of BPA During Pregnancy and Lactation

In humans and, more generally, primates, BPA-Gluc is undoubtedly the major metabolite of BPA produced in vivo. Because BPA-Gluc is not able to activate ERs, it is commonly accepted that the entero-hepatic conjugation is an efficient detoxification route, explaining both the rapid clearance of BPA from exposed organisms and BPA deactivation regarding its estrogenic properties (http://www.bisphenol-a.org 2010). However, the situation is more complex during perinatal life, in regards to an early exposure of the developing gut to BPA and the hypersensitivity of developing tissues to low levels of estrogens (Aksglaede et al. 2006). Of particular concern is the fact that the presence of free BPA has been repeatedly

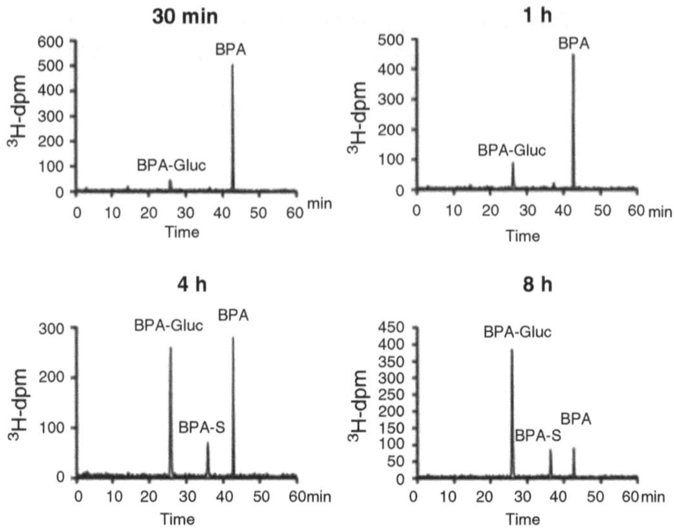

Fig. 2 Typical radio-chromatograms generated by HPLC showing the gradual biotransformation of ^{3}H-BPA into the corresponding metabolites BPA glucuronide (BPA-Gluc) and BPA sulphate (BPA-S), following the incubation of 0.1 μM BPA with human LS174T colonic cells at various time points after the beginning of incubations (Audebert et al., unpublished data)

demonstrated in umbilical cord blood, amniotic fluid and fetal plasma (Schonfelder et al. 2002; Yamada et al. 2002; Lee et al. 2008). These measurements revealed that a developing human fetus may be exposed to free BPA in the 1–3 ng/mL range (Vandenberg et al. 2007), which is the heart of the matter for risk assessment in humans (WHO 2010). Consistently, in both humans and rodents, BPA crosses the placental barrier and may accumulate in the fetus through umbilical blood and ingestion of amniotic fluid (Fig. 1).

Only a small number of studies in rats and mice have attempted to give a more complete picture of the distribution of BPA from mothers to fetuses and then to newborns during breastfeeding. In rats receiving a single oral dose of ^{14}C-BPA (500 μg/kg BW) at gestational day 18, radioactivity was found in the fetal intestine 24 h after maternal treatment (Kurebayashi et al. 2005). When the same oral dose was administered to lactating rats, most of the radioactivity found in newborns was also localized in the intestine after 24 h of breastfeeding (Kurebayashi et al. 2005). Using a subcutaneous route for BPA exposure with a low dose of ^{3}H-BPA (20 μg/kg BW), Zalko et al. (2003) reported that more than 4% of the dose was recovered in fetal tissues, with approximately 50% of BPA in active unconjugated forms 30 min after dosing. Of particular interest is the fact that most of the fetal residues of radiolabeled BPA are concentrated in maturating gut 24 h after treatment of dams, as illustrated in Fig. 3, an autoradiogram obtained from a pregnant mouse dosed at day 19 with ^{14}C-labeled BPA (Zalko et al., unpublished data).

Thus, independently from the route of exposure, the fetal gut is undoubtedly a target for BPA at late stages of gestation and during lactation. In support of free

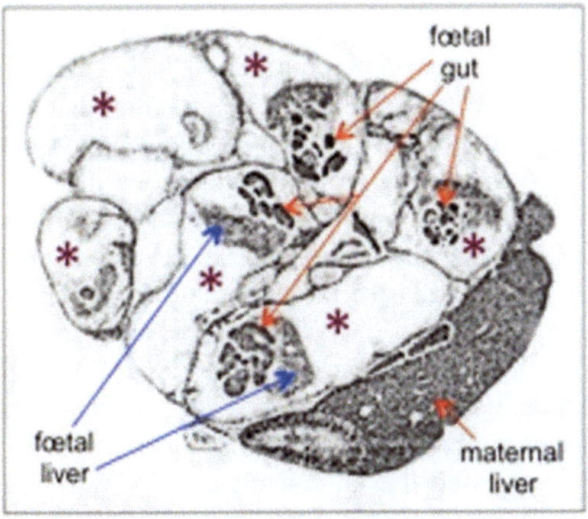

Fig. 3 Distribution of radiolabelled BPA residues in a pregnant mouse dosed with 2.8 mg ^{14}C-BPA at gestational day 17 (single dose) and euthanized 24 h after BPA dosage. The displayed autoradiogram, a transverse abdominal section through several fetal units (*asterisks*), demonstrates the presence of ^{14}C-BPA residues in the maternal liver and their specific accumulation in the maturating fotal liver and gut. Note that the gut is the organ where the most intense signal is observed (Zalko et al., unpublished data)

BPA with estrogenic effects in the maturing gut, recent data in pregnant rats indicated that BPA-Gluc from maternal blood could be de-conjugated back during its transfer to the fetus, at least partly, into the parent bio-active compound (Nishikawa et al. 2010). Additionally, authors showed that fetal liver displayed low UDPGT capability to metabolize BPA to BPA-Gluc, suggesting that active BPA might accumulate in the fetus at levels sufficient to influence gut development. It has also been postulated that free BPA in the sera of pregnant women could directly reach the fetus without a deactivation step during placental transport. Evidence for this hypothesis is supported by ex vivo perfusion of human placentas with free BPA, where only 3% of the transferred molecule in the fetal compartment was recovered in a conjugated form (Balakrishnan et al. 2010).

BPA and Intestinal Organogenesis: Lessons from *enopus laevis*

Currently published data suggest that the major targets of endocrine disruptors during development are primarily nuclear receptors such as ERs, as well as specific transport proteins. In both cases, a certain degree of structural homology between the xenobiotics and endogenous hormones is involved (Colborn 2004; Vandenberg et al. 2007). Both estrogens and the thyroid hormone T3 play a key role in embryonic and post-natal development in vertebrates (Morreale de Escobar 2001; Pepe et al. 2006), and BPA is also known for its antagonistic properties on thyroid hormone receptors (Moriyama et al. 2002). The context of postembryonic changes in amphibians is particularly interesting in determining the in vivo interaction pathways of BPA during development, because amphibian metamorphosis requires T3 and encompasses the postembryonic period in mammals, when T3 action is most critical (Heimeier et al. 2009). In *Xenopus laevis* tadpoles, Heimeier et al. (2009) showed that, after 4 days of exposure to the molecule, BPA (at doses consistent with the exposure levels of children during the first year of life) antagonized most of the genomic responses induced by T3, resulting in delayed larval metamorphosis with repercussions in the development of the intestine. It is possible that these effects occurred through a cross reaction between the ERs and T3 receptors, because estradiol exposure in early developmental stages (3–12 h post-egg fertilization), at similar concentrations (20 μM) of BPA, inhibits amphibian organogenesis, including abnormal gut coiling among other larval malformations (microcephaly, body length reduction, edema; Iwamuro et al. 2003). It should be stressed that in Xaenopus tadpoles, no conjugation of BPA occurs in these conditions (Fini et al. 2009); consequently, the observed effects can be attributed to parent BPA only.

Estrogen Receptors in Gut Epithelial Cells: A New Target for BPA

There is growing evidence that endogenous estrogens participate in the control of the intestinal barrier function in mammalians. Estrogen receptors are present in gut epithelial cells, mainly ERβ, expression of which is predominant in the human colon (Campbell-Thompson et al. 2001; Konstantinopoulos et al. 2003). Invalidation of ERβ in mice has revealed an important role of this signalling pathway on the architectural maintenance of the colonic epithelium. Epithelial cells are continuously renewed every 4–5 days (2–3 days in rodents) through a process initiated by stem cell division located in the bottom of the crypts, then differentiation toward the luminal surface before being eliminated by apoptosis and detachment (Yen and Wright 2006). In ERβ−/− mice, colonic epithelium is disorganized compared with wild-type littermates, showing enhanced proliferation of epithelial cells shifting towards the top of the crypt, as well as a defect in the formation of tight junctions (TJs) that control paracellular routes across the epithelial barrier (Wada-Hiraike et al. 2006). This first revealed the role of a "brake" for ERβ on epithelial renewal in the colon, in contrast to the proliferative effects of ERα (Barone et al. 2008), and likely a direct control on TJ permeability. Ultimately, an equilibrium between ERα− and β-signalling pathways in epithelial cells appears essential to maintain a functional barrier in the gut.

Because BPA has been shown to display a tenfold better affinity for ERβ, compared to ERα (Kuiper et al. 1997; Matthews et al. 2001), it has been hypothesized that free BPA in the gut lumen (among other xenoestrogens of concern in food) could directly influence the colonic epithelium. For instance, anti-proliferative effects have been reported in piglets orally exposed at birth to the phytoestrogen genistein (Chen et al. 2005), also a preferential activator of ERβ (Kuiper et al. 1998). Similar observations have been reported for the metalloestrogen cadmium, orally absorbed in ovariectomized (OVX) rats (Hofer et al. 2010). Consistently, recent observations by Houdeau and co-workers have demonstrated that low doses of BPA (5 μg/kg BW/day) administered to pregnant and lactating rats were effective in reducing epithelial cell proliferation in the colon of rat pups in the first week of life (unpublished data). Given the capability of BPA to cross the placental barrier in rodents (Zalko et al. 2003) as well as in humans (Balakrishnan et al. 2010), and because of its presence in the fetal gut (Fig. 3), anti-mitotic effects of BPA are likely to emerge at the end of pregnancy, a period wherein the epithelial turn-over is already fully active in the fetal intestine (Wagner et al. 2008).

In the epithelium of female rat colon, estradiol was also shown to reduce paracellular spaces through ERβ-mediated up-regulation of two transmembrane TJ proteins controlling the cell-cell interface, namely occludin and junctional adhesion molecule (JAM)-A (Braniste et al. 2009). This genomic effect appeared responsible for reproductive cycle-related fluctuations in colonic paracellular permeability in adult females, an effect that is due to natural variations in plasma estrogens (Braniste et al. 2009). When BPA was orally administered to OVX rats, a chronic and dose-dependent decrease in colonic paracellular permeability was observed, with a half maximal inhibitory dose tenfold below the current TDI of BPA (Braniste et al. 2010). Furthermore, a 24-h exposure of human colonic cells to BPA at concentrations consistent with those found in human fluid samples (0.1–10 nM) imposed a sharp decrease in paracellular permeability in cultured cell monolayers (Fig. 4). As reported in the rat, a permanent effect on human cells appeared closely related to an over-expression of occludin and JAM-A proteins at TJ apical sites, and involved ERβ ligand activity (Braniste et al. 2009, 2010). When chronic exposition occurs, it has been proposed that BPA interferes with fluid transport through the paracellular pathway, an important function of the colon that participates, together with the renal system, in the whole body fluid homeostasis and also maintains mucosal hydration (Masyuk et al. 2002; Geibel 2005). A permanent decrease in paracellular permeability induced by BPA may contribute to body fluid retention, a disorder commonly observed in high estrogenic states, as experienced by oral contraceptive users or in postmenopausal women with hormone replacement therapy (Oelkers 1996; Fruzzetti et al. 2007). In tissues such as the colon, where ERβ is the predominantly expressed ER, a potent tissue response to low doses of BPA may be related to a better coupling of ERβ with its co-activators (SRC-1e, TIF2 and TRAP220) in the presence of xenoestrogens, leading to transcriptional activity on reporter genes estimated to be 500 times greater for the BPA-ERβ complex, in comparison to a binding to ERα (Routledge et al. 2000; Welshons et al. 2006; Swedenborg et al. 2009).

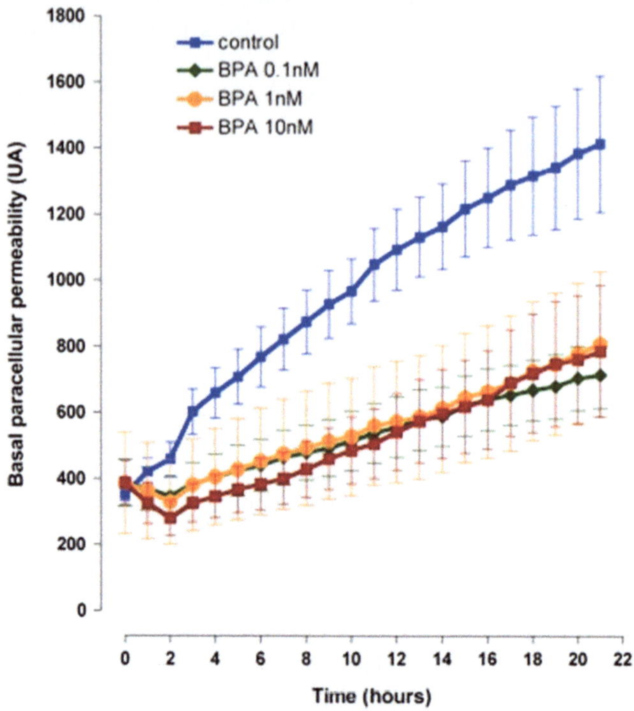

Fig. 4 Effect of BPA at concentrations detected in human fluids on paracellular permeability of human colonic cells. T84 cell monolayers (21 day old, differentiated cells) were incubated with 0.1–10 nM of BPA test samples and control (BPA vehicle) for 22 h. All treatments were performed in quadruplicate, and assay was repeated three times independently with similar results (Chaumaz and Houdeau, unpublished data)

BPA and Gut Development in Neonates: A Communication Breakdown that Persists in Later Life

The epithelial barrier is of prime importance in the maturing process of the gut. It is highly permeable at birth (Catassi et al. 1995), not solely for the absorption of nutrients necessary for body growth and fluid homeostasis of neonates but also to establish a facilitated crosstalk between bacteria that colonize the digestive tract and the immature immune cells located in the gut mucosa (Wagner et al. 2008; Dong et al. 2010). These maturing events in early life constitute the main elements for the development of tolerance and the recognition of pathogens and precede a period of "gut closure" to the external environment, to gain an effective barrier for life, with mature innate and adaptative immune regulatory mechanisms (Wagner et al. 2008; Turner 2009). Any alteration during this period is associated with a risk of developing functional bowel disorders and/or intestinal diseases in children and adults.

Recent literature emphasizes the hypersensitivity of the child to sex steroids, since circulating levels of estradiol in prepubertal children have appeared to be lower than originally claimed, and that estrogen-sensitive organs in young infancy may respond to exogenous sex steroids even at very low levels (Aksglaede et al. 2006). ERs are expressed in the human fetal gut (Brandenberger et al. 1997), with ERβ being the predominant ER detected in the colon (Takeyama et al. 2001). In perinatal life, early impacts of maternal estrogens on developing tissues and functions are normally limited by the presence of high-affinity steroid binding proteins in fetal blood, such as the alpha-fetoprotein (AFP), which restrict estrogen availability to target cells (Bakker and Baum 2008). However, studies performed on human plasma clearly indicate that BPA does not compete with the binding of estradiol to AFP (Milligan et al. 1998), which suggests that the xenoestrogen remains free in fetal blood and thus potentially active towards estradiol-sensitive tissue targets. Consistently, BPA at a low dose (5 μg/kg BW/day) orally administered to pregnant rats from mid-gestation until weaning of pups has been shown to decrease intestinal permeability in the offspring (Fig. 5). This effect

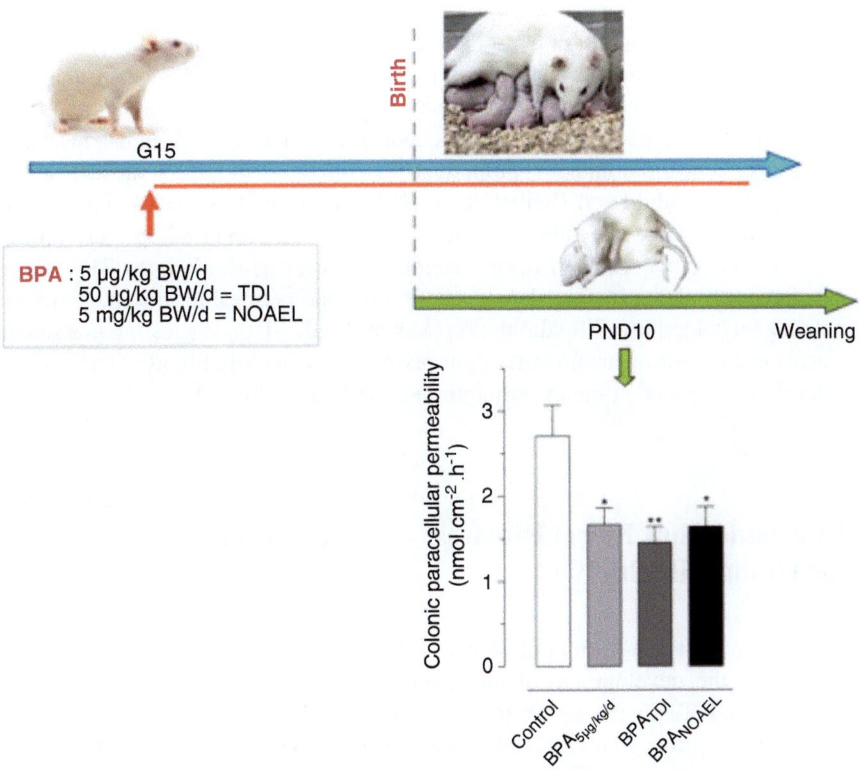

Fig. 5 Colonic paracellular permeability in 10-day-old rats exposed perinatally to BPA. Female rats were orally treated with different doses of BPA from midgestation (G15) until birth and throughout the lactation period. At 10 days post-natal (PND10), female pups were euthanized and the permeability was assessed on colonic strips mounted in Ussing chambers by measuring mucosal to serosal flux of FITC-labelled 4-kDa dextran over 1 h. *$p < 0.05$, **$p < 0.01$ vs. Control

persisted long after the end of BPA exposure, in female offspring only, and remained observable until adulthood when animals had been perinatally exposed to NOAEL doses through their mothers (Braniste et al. 2010). Furthermore, authors have suggested a profound influence on mucosal immune development, with persisting effects in later life, likely due to a chemically induced premature closure of the gut, impairing the luminal-to-mucosa crosstalk in early life. To provide this evidence, Braniste et al. (2010) reported an association between perinatal exposure to BPA and an imbalance of intestinal immune responses in experimental colitis in adulthood. These immune defects were characterized by enhanced neutrophil infiltration and colonic tissue damage in the inflamed area, and overproduction of T-Helper 1 pro-inflammatory cytokines, such as MIF, a preformed cytokine involved in the initiation and the maintenance of acute colonic inflammation (Houdeau et al. 2007). A developmental immunotoxicity has also been evidenced in mice prenatally exposed to BPA, showing enhanced basal levels of proinflammatory cytokines in blood in adulthood (Holladay et al. 2010), a decreased number of regulatory lymphocytes Treg cells involved in self-recognition, and the prevention of inflammatory autoimmune diseases (Yan et al. 2008). Interestingly, a recent study by Seibel et al. (2008) has also showed an abnormal inflammatory response in experimental colitis in adult rats exposed in utero and postnatally to the phytoestrogen genistein. It is thus noteworthy that, in contrast to the deleterious impact of a perinatal exposure, both BPA and genistein exhibit protective, anti-inflammatory properties in colitis when oral exposure occurs directly during adulthood (Seibel et al. 2009; Braniste et al. 2010). To date, the reasons for an age-dependent variation in genistein impacts on gut immune responses have not received further attention. However, similarly to BPA, recent data have revealed a sustained decrease of intestinal permeability in rats fed a diet rich in genistein, due to ER stimulation (Moussa et al. 2010), suggesting that several classes of xenoestrogens impairing gut permeability in early life may have similar deleterious impacts on the developing immune barrier (Fig. 6).

BPA and Adult Bowel Diseases: A Serious Concern for Human Health?

The notion that the hormonal status and exogenous estrogens in women may predispose the development of an inflammatory bowel disease (IBD) has often been debated (El-Tawil 2008). Recent clinical studies have associated long-term use of oral contraceptives or hormone replacement therapy with the risk of developing Crohn's disease (CD) or ulcerative colitis (UC), but a causal relation is still controversial (Cosnes et al. 1999; Garcia Rodriguez et al. 2005; Kane and Reddy 2008). In humans, both CD and UC are intestinal pathologies characterized by alternating bursts of activity and periods of remission. While the etiology remains unknown, between genetic predisposition (mutation of NOD2, CARD15 involved

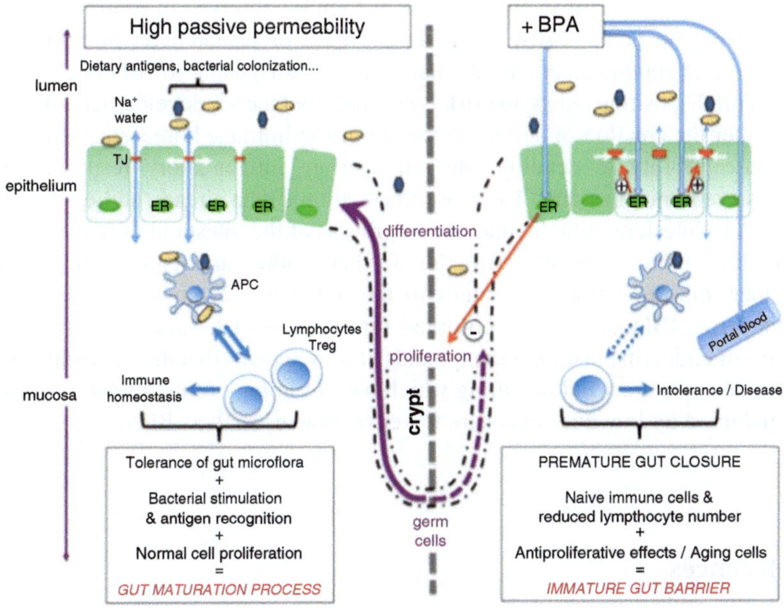

Fig. 6 Intestinal barrier maturation and immune cell development in the newborn. The gut epithelium is composed of a single layer of columnar cells that proliferate from the germ cells located at the bottom of intestinal crypts. Epithelial cells are renewed permanently (≈96 h) for life, beginning from the fetal stages. Epithelial permeability between cells (paracellular transport) is determined by apical tight junctions (TJs) that seal paracellular spaces. Intestinal permeability is permissive at birth and during the first weeks/months of life to allow easy exchanges between luminal content (colonizing bacteria, toxins, dietary antigens…) and mucosal immune cells (present at birth but immature). Antigen presenting cells (APC: dendritic cells, M cells) sample foreign materials and deliver these sensors to regulatory T cells (Treg) that orchestrate the response in normal immune tolerance of the microflora and recognition of various antigens (homeostasis). A perinatal exposure to BPA, as a ligand of estrogen receptors (ERs) in epithelial cells, limits epithelial cell proliferation and decreases paracellular transport through the stimulation of TJ protein expression. This premature gut closure allows reduction in the passage of bacterial sensors and other luminal signals across the epithelium and impairs education of underlying immune system (dysregulation of immune homeostasis, digestive disease susceptibility). A transcellular transport of bacterial products (through Peyer's patches, Toll like receptors of surface) is not represented in this schema

in bacteria recognition) and environmental factors (antibiotics, anti-inflammatory drugs) (Mayer 2010; Shaw et al. 2010), currently available data suggest a more conventional break in immune homeostasis of the intestine, as the result of an imbalance between the progressive colonization by the commensal flora and host defenses, and the occurrence of an acute and uncontrolled inflammatory response (Shih and Targan 2009; Mayer 2010). While animal studies mostly support the protective effects of estradiol and xenoestrogens in the intestinal inflammatory response, these effects are mainly observed in adults exposed chronically to these molecules (Harnish et al. 2004; Houdeau et al. 2007; Seibel et al. 2009; Braniste et al. 2010). The demonstration that perinatal exposure to xenoestrogens

(BPA, genistein) results in opposite effects to those observed in adults (Seibel et al. 2008; Braniste et al. 2010) opens a new field of research on the potential of endocrine disruptors as environmental factors predisposing to IBD later in life. If such a predisposition exists, the risk is not the spontaneous development of diseases but rather the creation of an imbalance at birth in immune homeostasis, increasing the susceptibility to develop later an aberrant inflammatory response against aggressive luminal assault. The most recent animal data already address the ability of BPA to interfere with the maturation process of the intestinal immune system in early life (Braniste et al. 2010), but further studies are required to determine whether this disruption contributes to the imbalance in immune responses that characterize IBD. As already observed for many other tissular and physiological targets of endocrine disruptor compounds, it now appears that the perinatal period is a critical exposure window during which the physiology of the maturating gut can be disrupted by low-dose exposure to xenoestrogens such as BPA.

References

Aksglaede L, Juul A, Leffers H, Skakkebaek NE, Andersson AM (2006) The sensitivity of the child to sex steroids: possible impact of exogenous estrogens. Hum Reprod Update 12:341–349

Bakker J, Baum MJ (2008) Role for estradiol in female-typical brain and behavioral sexual differentiation. Front Neuroendocrinol 29:1–16

Balakrishnan B, Henare K, Thorstensen EB, Ponnampalam AP, Mitchell MD (2010) Transfer of bisphenol A across the human placenta. Am J Obstet Gynecol 202(393):e391–e397

Barone M, Tanzi S, Lofano K, Scavo MP, Guido R, Demarinis L, Principi MB, Bucci A, Di Leo A (2008) Estrogens, phytoestrogens and colorectal neoproliferative lesions. Genes Nutr 3:7–13

Bondesson M, Jonsson J, Pongratz I, Olea N, Cravedi JP, Zalko D, Hakansson H, Halldin K, Di Lorenzo D, Behl C, Manthey D, Balaguer P, Demeneix B, Fini JB, Laudet V, Gustafsson JA (2009) A CASCADE of effects of bisphenol A. Reprod Toxicol 28:563–567

Borghoff SJ, Birnbaum LS (1985) Age-related changes in glucuronidation and deglucuronidation in liver, small intestine, lung, and kidney of male Fischer rats. Drug Metab Dispos 13:62–67

Brandenberger AW, Tee MK, Lee JY, Chao V, Jaffe RB (1997) Tissue distribution of estrogen receptors alpha (ER-alpha) and beta (ER-beta) mRNA in the midgestational human fetus. J Clin Endocrinol Metab 82:3509–3512

Braniste V, Leveque M, Buisson-Brenac C, Bueno L, Fioramonti J, Houdeau E (2009) Oestradiol decreases colonic permeability through oestrogen receptor beta-mediated up-regulation of occludin and junctional adhesion molecule-A in epithelial cells. J Physiol 587(Pt 13):3317–3328

Braniste V, Jouault A, Gaultier E, Polizzi A, Buisson-Brenac C, Leveque M, Martin PG, Theodorou V, Fioramonti J, Houdeau E (2010) Impact of oral bisphenol A at reference doses on intestinal barrier function and sex differences after perinatal exposure in rats. Proc Natl Acad Sci USA 107:448–453

Brede C, Fjeldal P, Skjevrak I, Herikstad H (2003) Increased migration levels of bisphenol A from polycarbonate baby bottles after dishwashing, boiling and brushing. Food Addit Contam 20:684–689

Brotons JA, Olea-Serrano MF, Villalobos M, Pedraza V, Olea N (1995) Xenoestrogens released from lacquer coatings in food cans. Environ Health Perspect 103:608–612

Burridge E (2003) Bisphenol A product profile. Eur Chem News 17:14–20

Calafat AM, Kuklenyik Z, Reidy JA, Caudill SP, Ekong J, Needham LL (2005) Urinary concentrations of bisphenol A and 4-nonylphenol in a human reference population. Environ Health Perspect 113:391–395

Calafat AM, Ye X, Wong LY, Reidy JA, Needham LL (2008) Exposure of the U.S. population to bisphenol A and 4-tertiary-octylphenol: 2003–2004. Environ Health Perspect 116:39–44

Campbell-Thompson M, Lynch IJ, Bhardwaj B (2001) Expression of estrogen receptor (ER) subtypes and ERbeta isoforms in colon cancer. Cancer Res 61:632–640

Cao XL, Corriveau J, Popovic S (2010) Sources of low concentrations of bisphenol A in canned beverage products. J Food Prot 73:1548–1551

Carwile JL, Luu HT, Bassett LS, Driscoll DA, Yuan C, Chang JY, Ye X, Calafat AM, Michels KB (2009) Polycarbonate bottle use and urinary bisphenol A concentrations. Environ Health Perspect 117:1368–1372

Catassi C, Bonucci A, Coppa GV, Carlucci A, Giorgi PL (1995) Intestinal permeability changes during the first month: effect of natural versus artificial feeding. J Pediatr Gastroenterol Nutr 21:383–386

Chen AC, Berhow MA, Tappenden KA, Donovan SM (2005) Genistein inhibits intestinal cell proliferation in piglets. Pediatr Res 57:192–200

Colborn T (2004) Neurodevelopment and endocrine disruption. Environ Health Perspect 112:944–949

Cosnes J, Carbonnel F, Carrat F, Beaugerie L, Gendre JP (1999) Oral contraceptive use and the clinical course of Crohn's disease: a prospective cohort study. Gut 45:218–222

Dong P, Yang Y, Wang WP (2010) The role of intestinal bifidobacteria on immune system development in young rats. Early Hum Dev 86:51–58

El-Tawil AM (2008) Oestrogens and Crohn's disease: the missed link. Andrologia 40:141–145

Emmerson E, Campbell L, Ashcroft GS, Hardman MJ (2009) Unique and synergistic roles for 17beta-estradiol and macrophage migration inhibitory factor during cutaneous wound closure are cell type specific. Endocrinology 150:2749–2757

Enmark E, Pelto-Huikko M, Grandien K, Lagercrantz S, Lagercrantz J, Fried G, Nordenskjold M, Gustafsson JA (1997) Human estrogen receptor beta-gene structure, chromosomal localization, and expression pattern. J Clin Endocrinol Metab 82:4258–4265

Fini JB, Dolo L, Cravedi JP, Demeneix B, Zalko D (2009) Metabolism of the endocrine disruptor BPA by *Xenopus laevis* tadpoles. Ann NY Acad Sci 1163:394–397

Fruzzetti F, Lazzarini V, Ricci C, Quirici B, Gambacciani M, Paoletti AM, Genazzani AR (2007) Effect of an oral contraceptive containing 30 microg ethinylestradiol plus 3 mg drospirenone on body composition of young women affected by premenstrual syndrome with symptoms of water retention. Contraception 76:190–194

Garcia Rodriguez LA, Gonzalez-Perez A, Johansson S, Wallander MA (2005) Risk factors for inflammatory bowel disease in the general population. Aliment Pharmacol Ther 22:309–315

Geibel JP (2005) Secretion and absorption by colonic crypts. Annu Rev Physiol 67:471–490

Harnish DC, Albert LM, Leathurby Y, Eckert AM, Ciarletta A, Kasaian M, Keith JC Jr (2004) Beneficial effects of estrogen treatment in the HLA-B27 transgenic rat model of inflammatory bowel disease. Am J Physiol Gastrointest Liver Physiol 286:G118–G125

Heimeier RA, Das B, Buchholz DR, Shi YB (2009) The xenoestrogen bisphenol A inhibits postembryonic vertebrate development by antagonizing gene regulation by thyroid hormone. Endocrinology 150:2964–2973

Heitkemper MM, Cain KC, Jarrett ME, Burr RL, Hertig V, Bond EF (2003) Symptoms across the menstrual cycle in women with irritable bowel syndrome. Am J Gastroenterol 98:420–430

Hofer N, Diel P, Wittsiepe J, Wilhelm M, Kluxen FM, Degen GH (2010) Investigations on the estrogenic activity of the metallohormone cadmium in the rat intestine. Arch Toxicol 84:541–552

Holladay SD, Xiao S, Diao H, Barber J, Nagy T, Ye X, Gogal RM Jr (2010) Perinatal bisphenol A exposure in C57B6/129svj male mice: potential altered cytokine/chemokine production in adulthood. Int J Environ Res Public Health 7:2845–2852

Houdeau E, Moriez R, Leveque M, Salvador-Cartier C, Waget A, Leng L, Bueno L, Bucala R, Fioramonti J (2007) Sex steroid regulation of macrophage migration inhibitory factor in normal and inflamed colon in the female rat. Gastroenterology 132:982–993

Howe SR, Borodinsky L (1998) Potential exposure to bisphenol A from food-contact use of polycarbonate resins. Food Addit Contam 15:370–375

Inoue K, Kato K, Yoshimura Y, Makino T, Nakazawa H (2000) Determination of bisphenol A in human serum by high-performance liquid chromatography with multi-electrode electrochemical detection. J Chromatogr B Biomed Sci Appl 749:17–23

Inoue H, Yuki G, Yokota H, Kato S (2003) Bisphenol A glucuronidation and absorption in rat intestine. Drug Metab Dispos 31:140–144

Iwamuro S, Sakakibara M, Terao M, Ozawa A, Kurobe C, Shigeura T, Kato M, Kikuyama S (2003) Teratogenic and anti-metamorphic effects of bisphenol A on embryonic and larval *Xenopus laevis*. Gen Comp Endocrinol 133:189–198

Kane SV, Reddy D (2008) Hormonal replacement therapy after menopause is protective of disease activity in women with inflammatory bowel disease. Am J Gastroenterol 103:1193–1196

Kim HS, Han SY, Yoo SD, Lee BM, Park KL (2001) Potential estrogenic effects of bisphenol-A estimated by in vitro and in vivo combination assays. J Toxicol Sci 26:111–118

Konstantinopoulos PA, Kominea A, Vandoros G, Sykiotis GP, Andricopoulos P, Varakis I, Sotiropoulou-Bonikou G, Papavassiliou AG (2003) Oestrogen receptor beta (ERbeta) is abundantly expressed in normal colonic mucosa, but declines in colon adenocarcinoma paralleling the tumour's dedifferentiation. Eur J Cancer 39:1251–1258

Kuiper GG, Carlsson B, Grandien K, Enmark E, Haggblad J, Nilsson S, Gustafsson JA (1997) Comparison of the ligand binding specificity and transcript tissue distribution of estrogen receptors alpha and beta. Endocrinology 138:863–870

Kuiper GG, Lemmen JG, Carlsson B, Corton JC, Safe SH, van der Saag PT, van der Burg B, Gustafsson JA (1998) Interaction of estrogenic chemicals and phytoestrogens with estrogen receptor beta. Endocrinology 139:4252–4263

Kurebayashi H, Nagatsuka S, Nemoto H, Noguchi H, Ohno Y (2005) Disposition of low doses of 14C-bisphenol A in male, female, pregnant, fetal, and neonatal rats. Arch Toxicol 79:243–252

Lee YJ, Ryu HY, Kim HK, Min CS, Lee JH, Kim E, Nam BH, Park JH, Jung JY, Jang DD, Park EY, Lee KH, Ma JY, Won HS, Im MW, Leem JH, Hong YC, Yoon HS (2008) Maternal and fetal exposure to bisphenol A in Korea. Reprod Toxicol 25:413–419

Lopez-Cervantes J, Paseiro-Losada P (2003) Determination of bisphenol A in, and its migration from, PVC stretch film used for food packaging. Food Addit Contam 20:596–606

Masyuk AI, Marinelli RA, LaRusso NF (2002) Water transport by epithelia of the digestive tract. Gastroenterology 122:545–562

Matthews JB, Twomey K, Zacharewski TR (2001) In vitro and in vivo interactions of bisphenol A and its metabolite, bisphenol A glucuronide, with estrogen receptors alpha and beta. Chem Res Toxicol 14:149–157

Mayer L (2010) Evolving paradigms in the pathogenesis of IBD. J Gastroenterol 45:9–16

Milligan SR, Khan O, Nash M (1998) Competitive binding of xenobiotic oestrogens to rat alpha-fetoprotein and to sex steroid binding proteins in human and rainbow trout (*Oncorhynchus mykiss*) plasma. Gen Comp Endocrinol 112:89–95

Moriyama K, Tagami T, Akamizu T, Usui T, Saijo M, Kanamoto N, Hataya Y, Shimatsu A, Kuzuya H, Nakao K (2002) Thyroid hormone action is disrupted by bisphenol A as an antagonist. J Clin Endocrinol Metab 87:5185–5190

Morreale de Escobar G (2001) The role of thyroid hormone in fetal neurodevelopment. J Pediatr Endocrinol Metab 14(Suppl 6):1453–1462

Moussa L, Grimaldi C, Paul F, Braniste V, Tondereau V, Eutamene H, Bueno L, Fioramonti J, Houdeau E, Theodorou V (2010) Protective effect of a phytoestrogen-enriched diet on the increase in visceral sensitivity and intestinal permeability induced by acute stress in female rats. Gastroenterology 138:S27

Nam SH, Seo YM, Kim MG (2010) Bisphenol A migration from polycarbonate baby bottle with repeated use. Chemosphere 79:949–952

Nishikawa M, Iwano H, Yanagisawa R, Koike N, Inoue H, Yokota H (2010) Placental transfer of conjugated bisphenol A and subsequent reactivation in the rat fetus. Environ Health Perspect 118:1196–1203

O'Leary KA, Day AJ, Needs PW, Mellon FA, O'Brien NM, Williamson G (2003) Metabolism of quercetin-7- and quercetin-3-glucuronides by an in vitro hepatic model: the role of human beta-glucuronidase, sulfotransferase, catechol-O-methyltransferase and multi-resistant protein 2 (MRP2) in flavonoid metabolism. Biochem Pharmacol 65:479–491

O'Mahony F, Harvey BJ (2008) Sex and estrous cycle-dependent rapid protein kinase signaling actions of estrogen in distal colonic cells. Steroids 73(9–10):889–894

Oelkers WK (1996) Effects of estrogens and progestogens on the renin-aldosterone system and blood pressure. Steroids 61:166–171

Olea N, Pulgar R, Perez P, Olea-Serrano F, Rivas A, Novillo-Fertrell A, Pedraza V, Soto AM, Sonnenschein C (1996) Estrogenicity of resin-based composites and sealants used in dentistry. Environ Health Perspect 104:298–305

Pepe GJ, Burch MG, Albrecht ED (2006) Developmental regulation of the sodium/hydrogen ion exchangers and their regulatory factors in baboon placental syncytiotrophoblast. Endocrinology 147:2986–2996

Pfaffl MW, Lange IG, Meyer HH (2003) The gastrointestinal tract as target of steroid hormone action: quantification of steroid receptor mRNA expression (AR, ERalpha, ERbeta and PR) in 10 bovine gastrointestinal tract compartments by kinetic RT-PCR. J Steroid Biochem Mol Biol 84:159–166

Routledge EJ, White R, Parker MG, Sumpter JP (2000) Differential effects of xenoestrogens on coactivator recruitment by estrogen receptor (ER) alpha and ERbeta. J Biol Chem 275:35986–35993

Sajiki J, Yonekubo J (2004) Leaching of bisphenol A (BPA) from polycarbonate plastic to water containing amino acids and its degradation by radical oxygen species. Chemosphere 55:861–867

Schonfelder G, Wittfoht W, Hopp H, Talsness CE, Paul M, Chahoud I (2002) Parent bisphenol A accumulation in the human maternal-fetal-placental unit. Environ Health Perspect 110: A703–A707

Seibel J, Molzberger AF, Hertrampf T, Laudenbach-Leschowski U, Degen GH, Diel P (2008) In utero and postnatal exposure to a phytoestrogen-enriched diet increases parameters of acute inflammation in a rat model of TNBS-induced colitis. Arch Toxicol 82:941–950

Seibel J, Molzberger AF, Hertrampf T, Laudenbach-Leschowski U, Diel P (2009) Oral treatment with genistein reduces the expression of molecular and biochemical markers of inflammation in a rat model of chronic TNBS-induced colitis. Eur J Nutr 48:213–220

Shaw SY, Blanchard JF, Bernstein CN (2010) Association between the use of antibiotics in the first year of life and pediatric inflammatory bowel disease. Am J Gastroenterol 105:2687–2692

Shih DQ, Targan SR (2009) Insights into IBD pathogenesis. Curr Gastroenterol Rep 11:473–480

Sperker B, Backman JT, Kroemer HK (1997) The role of beta-glucuronidase in drug disposition and drug targeting in humans. Clin Pharmacokinet 33:18–31

Speyer CL, Rancilio NJ, McClintock SD, Crawford JD, Gao H, Sarma JV, Ward PA (2005) Regulatory effects of estrogen on acute lung inflammation in mice. Am J Physiol Cell Physiol 288:C881–C890

Swedenborg E, Ruegg J, Makela S, Pongratz I (2009) Endocrine disruptive chemicals: mechanisms of action and involvement in metabolic disorders. J Mol Endocrinol 43:1–10

Takeyama J, Suzuki T, Inoue S, Kaneko C, Nagura H, Harada N, Sasano H (2001) Expression and cellular localization of estrogen receptors alpha and beta in the human fetus. J Clin Endocrinol Metab 86:2258–2262

Turner JR (2009) Intestinal mucosal barrier function in health and disease. Nat Rev Immunol 9:799–809

Vandenberg LN, Hauser R, Marcus M, Olea N, Welshons WV (2007) Human exposure to bisphenol A (BPA). Reprod Toxicol 24:139–177
Verdu EF, Deng Y, Bercik P, Collins SM (2002) Modulatory effects of estrogen in two murine models of experimental colitis. Am J Physiol Gastrointest Liver Physiol 283:G27–G36
Wada-Hiraike O, Imamov O, Hiraike H, Hultenby K, Schwend T, Omoto Y, Warner M, Gustafsson JA (2006) Role of estrogen receptor beta in colonic epithelium. Proc Natl Acad Sci USA 103:2959–2964
Wagner CL, Taylor SN, Johnson D (2008) Host factors in amniotic fluid and breast milk that contribute to gut maturation. Clin Rev Allergy Immunol 34:191–204
Welshons WV, Nagel SC, vom Saal FS (2006) Large effects from small exposures. III. Endocrine mechanisms mediating effects of bisphenol A at levels of human exposure. Endocrinology 147(6 Suppl):S56–S69
WHO (2010) Joint FAO/WHO expert meeting to review toxicological and health aspects of bisphenol A, Ottawa, Canada
Wilson ME, Westberry JM, Prewitt AK (2008) Dynamic regulation of estrogen receptor-alpha gene expression in the brain: a role for promoter methylation? Front Neuroendocrinol 29:375–385
Yamada H, Furuta I, Kato EH, Kataoka S, Usuki Y, Kobashi G, Sata F, Kishi R, Fujimoto S (2002) Maternal serum and amniotic fluid bisphenol A concentrations in the early second trimester. Reprod Toxicol 16:735–739
Yan H, Takamoto M, Sugane K (2008) Exposure to Bisphenol A prenatally or in adulthood promotes T(H)2 cytokine production associated with reduction of CD4CD25 regulatory T cells. Environ Health Perspect 116:514–519
Yen TH, Wright NA (2006) The gastrointestinal tract stem cell niche. Stem Cell Rev 2:203–212
Zalko D, Soto AM, Dolo L, Dorio C, Rathahao E, Debrauwer L, Faure R, Cravedi JP (2003) Biotransformations of bisphenol A in a mammalian model: answers and new questions raised by low-dose metabolic fate studies in pregnant CD1 mice. Environ Health Perspect 111:309–319

Adverse Trends of Male Reproductive Health in Two Nordic Countries Indicate Environmental Problems

Jorma Toppari and Niels E. Skakkebaek

Abstract Current male reproductive health is poor and infertility will be an increasing problem. The reasons are not known, but emerging evidence from experimental animal studies, wildlife, and epidemiological studies combined to our biological and clinical understanding support the hypothesis that environmental endocrine disrupters contribute to the problem. It is therefore important to increase awareness of male reproductive health problems and strengthen interdisciplinary, translational research in this field as specified recently by European Science Foundation (ESF; ESF Science Policy Briefing, September 2010, www.esf.org). Long-term epidemiological studies combining genetic and environmental information with good clinical examinations, biological pathway analyses and effective bioinformatics will be crucial to understand the etiology of poor male reproductive health. Targeted research efforts to better understand mechanisms involved in these processes, including late effects of fetal exposure, will provide paths forward for better male reproductive health.

Introduction

Fertility has been one of the highest valued assets in human life since the biblical era. The recognition of the invention of in vitro fertilization (IVF) techniques by the awarding of the Nobel Prize in 2010 showed that our values have not changed much over millennia. However, since the 1960s, the main focus in reproductive medicine has changed from the search for reasons of infertility to the development of contraceptives and later on to studies of new techniques of assisted reproduction, such as IVF and intracytoplasmic sperm injection (ICSI). The pharmaceutical industry guided a lot of the funding for contraceptive studies and fertility treatment

---

J. Toppari (✉)
Departments of Physiology and Paediatrics, University of Turku, Turku, Finland
e-mail: jorma.toppari@utu.fi

J.-P. Bourguignon et al. (eds.), *Multi-System Endocrine Disruption*,
Research and Perspectives in Endocrine Interactions,
DOI 10.1007/978-3-642-22775-2_10, © Springer-Verlag Berlin Heidelberg 2011

trials that occupied the research time of busy clinicians. It is no wonder that, in this research atmosphere, results of deteriorating semen quality and declining fecundity were met with scepticism among researchers who invested all their efforts into either contraception or fertility treatments. The latter also attracted several reproductive scientists into private enterprises. It is a major advance that we have good treatment options for infertility and the best possible caretakers providing those treatments; on the other hand, we cannot emphasize enough the value of the most important goal, which is the prevention of infertility.

Fertility rates have declined dramatically over a short time period covering only one or two generations. The introduction of modern birth control methods and numerous societal factors have played major roles in this change. However, certain biological factors need to be present in order for conception to occur (Skakkebæk et al. 2006). The general postponement of first childbirth by modern women has caused a shift in childbearing into an age range in which fecundity may be decreasing. Furthermore, adverse trends in male reproductive health may be behind the widespread need for assisted reproductive techniques (ART). The burning research question is, what is the reason for the many reproductive health problems, such as the increased incidence of testis cancer and poor semen quality? A comparison of countries where testis cancer incidence is low and semen quality is good with countries with the opposite disease pattern offers an epidemiological opportunity to address the question. We have used this opportunity to compare the development and function of the male reproductive system in Denmark and Finland, neighboring countries with poor and good reproductive health, respectively.

Testicular Gem Cell Cancer (TGCC)

Over the past decades we have witnessed a steady increase in TGCC in industrialized countries, particularly among Caucasians (Fig. 1). Although there are clear genetic aspects of this disease (Kanetsky et al. 2009; Rapley et al. 2009; Goriely et al. 2009), the trends in incidence document a strong environmental component. In Denmark, where the cancer registry was established in 1943, the incidence rates have quadrupled. Similar trends have been seen in other countries, although the current incidence rates vary enormously.

The differences in rates of TGCC between Denmark and Finland are quite remarkable (Bray et al. 2006a, b). During the 1970s, the incidence was four times higher in Denmark compared with Finland. Interestingly, there has recently been a relatively sharp increase in Finland (Jørgensen et al. 2011) whereas the Danish rates may have stabilized at the high levels seen around 2000. The Danish-Finnish pattern in TGCC was, in fact, what inspired us to do the large Danish-Finnish studies on male reproductive health mentioned below.

During the period of rising rates in TGCC, epidemiologists noted that there was a birth cohort effect that was stronger than the time effect: the men who were born more recently had a particularly high rate of TGCC. Such findings suggest that

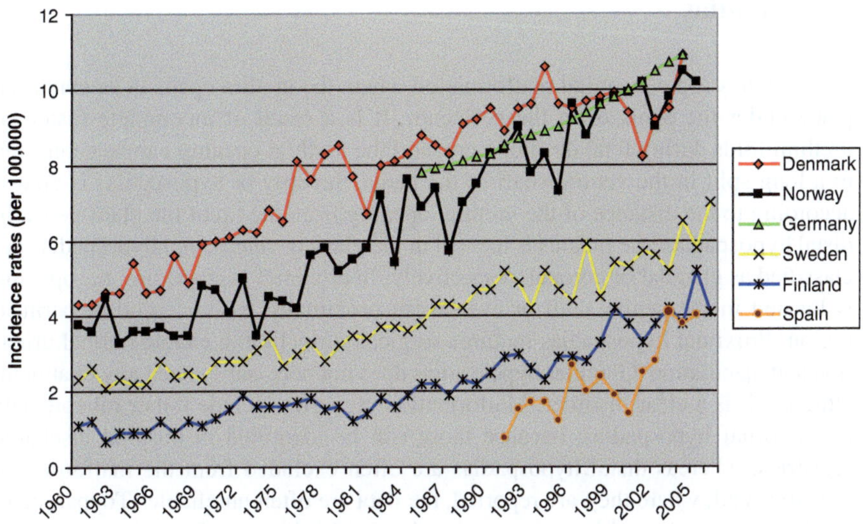

Fig. 1 Trends in incidence rates of testicular cancer in Nordic countries – Denmark, Finland, Norway, and Sweden – and in Germany and Spain. Data were collected from "Association of Nordic Cancer Registries," "Cancer in Germany" and Llanes et al. (2008). Courtesy of N. Jørgensen

perinatal events may play a role. Other epidemiological studies (Bergström et al. 1996) also suggested that environmental factors around birth were important: the Danish, Norwegian and Swedish boys born during World War II had lower risks of TGCC in adulthood than estimated from trends. It is now well established that TGCC develops from carcinoma in situ testis, which is a cell pattern consisting of gonocyte-like germ cells that express pluripotency genes, including C-Kit, NANOG, Oct-4, AP2gamma and several others. (Rajpert-De Meyts 2006). Thus basic and epidemiological studies all point to an early fetal origin of TGCC.

In most cases of TGCC, there is no obvious cause. However, mutations causing abnormal sexual development with severe dysgenesis of the gonad are well known to increase the risk of TGCC (including a variant of TGCC called gonadoblastoma). Although such cases are extremely rare, they have enlightened our thinking about the biology of TGCC. Dysgenesis of the testis, although in modest form, can also be seen in testicular tissue from men with cryptorchidism, infertility and hypospadias. These observations of dysgenesis, together with epidemiological data showing that TGCC, hypospadias, poor semen quality and cryptorchidism are risk factors for each other, lead to the possibility of a testicular dysgenesis syndrome (TDS) that may exhibit all or some of these symptoms.

Thus, there is a strong environmental component in the trends in TGCC. In addition, available data suggest that factors operating in early life, perhaps even as early as in the first trimester, are important. Therefore, research is now focusing on the possibility that exposures of pregnant women could result in TGCC (Hardell et al. 2003, 2006).

Hypospadias

Hypospadias is a congenital malformation where the urethra opens in an abnormal place under the penis or in the perineum. It is a result of incomplete fusion of urethral folds during fetal development, and the urethral opening can be connected to a long split in the ventral shaft of the penis. Severity of hypospadias is graded according to the distance of the urethral opening from the tip of the glans penis. In distal hypospadias, the meatus is located in the glans or sulcus, and hypospadias are classified as glanular or coronal, respectively. In proximal hypospadias, the opening is located in the penile shaft or even more proximally in the scrotal or perineal region. Proximal hypospadias requires surgical correction to enable normal urination and ejaculation. Glanular hypospadias does not necessarily need any treatment, although it is a clear disorder. Malformation registries include rather reliable data on proximal hypospadias, because those can be identified in hospital discharge registries, whereas distal hypospadias are often excluded from the malformation registries and, where they are reported, the data are often unreliable (Toppari et al. 2001). Nevertheless, increasing trends in the birth rate of hypospadias were reported in several European countries until 1980s and in USA until 1990s on the basis of registry data. Large regional differences were also apparent. We performed a prospective joint cohort study in Denmark and Finland to assess the prevalence of hypospadias with uniform diagnostic criteria (Virtanen et al. 2001; Boisen et al. 2005; Fig. 2). The birth rate of hypospadias in Denmark (1.03%) was threefold higher than in Finland (0.27%); only one case of proximal hypospadias was identified in the Finnish cohort (Virtanen et al. 2001), whereas the Danish cohort included several (Boisen et al. 2005). The Finnish figures in the cohort study were a little higher than those reported from previous Finnish studies (Aho et al. 2000). A Dutch study from the same period reported a prevalence that was almost as high as in the Danish study (Pierik et al. 2002). A large registry-based Danish study has

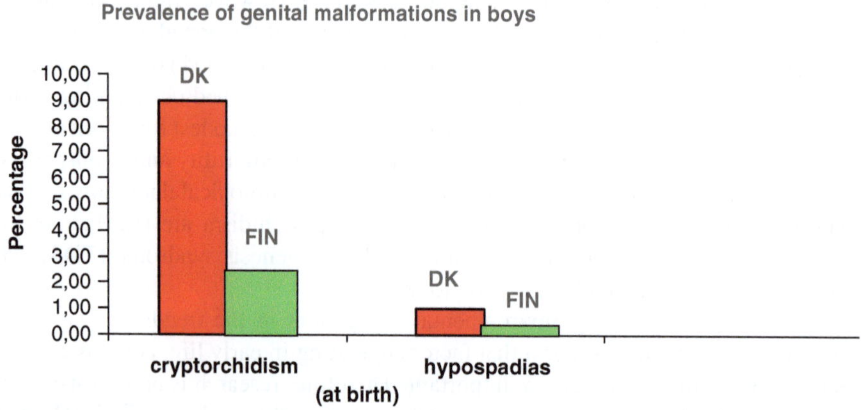

Fig. 2 Birth rates of cryptorchidism and hypospadias in Denmark and Finland (Boisen et al. 2004, 2005)

shown an increase in the rate of hypospadias continuing to the end of the observation period (Lund et al. 2009).

Androgen insensitivity or defects in androgen biosynthesis cause hypospadias (Quigley et al. 1995). However, endocrine abnormalities can be found in only about 15% of the cases (Rey et al. 2005). Genetic polymorphisms have also been associated with the risk of hypospadias, and the most recent finding on common variants in diacylglycerol kinase κ (DGGK) provided the strongest association so far (van der Zanden et al. 2011a, b). The study did not explain how the gene variation caused the effect. In animal experiments, fetal exposure to anti-androgens and/or estrogens causes hypospadias even at low levels of exposure when several compounds affecting the same signalling pathway are combined (Rider et al. 2008; Christiansen et al. 2008; Hass et al. 2007). Therefore we have hypothesized that endocrine disrupters with anti-androgenic or estrogenic properties may have contributed to the adverse trends in hypospadias rates. Testing the hypothesis in an epidemiological setting would require collection of very large numbers of cases and controls, which is difficult, because hypospadias is still rare albeit common compared to most other birth defects. However, it is closely related to another birth defect, cryptorchidism, which is ten times more common. Therefore, we can study the association of cryptorchidism with exposures and then ask specifically whether positive associations can also be found for hypospadias.

Cryptorchidism

Testes descend to the bottom of the scrotum during the last trimester of gestation. When one or both of the testes fail to descend normally, the boy has congenital cryptorchidism. The testes can be found anywhere on the normal route of descent, and the severity of the condition is classified according to the testicular position from most difficult to mildest in the following order: abdominal, inguinal, suprascrotal, and high scrotal. Sometimes the testis can move freely up and down, which is called retractile or sliding testis and is considered normal. If the testis is found outside of the normal descent route, it is called ectopic. Testicular descent is regulated by the Leydig cell-derived hormones testosterone and insulin-like peptide 3 (INSL3), and Leydig cell function is stimulated by the pituitary luteinizing hormone (LH). In cryptorchid boys, LH levels are slightly elevated at 3 months of age, suggesting that Leydig cell function is somewhat compromised (Suomi et al. 2006). Thus, cryptorchidism is a sign of testicular dysfunction. Later on, cryptorchidism is associated with impaired semen quality (Taskinen et al. 1997) and subfertility (Lee 1993, 2005). Early orchidopexy prevents further damage to the germ cells, which are dependent on testicular temperature (Kollin et al. 2007; Taskinen et al. 1997), but it does not correct the whole problem (Hadziselimovic et al. 2007). Cryptorchidism is also a risk factor for testis cancer, increasing the risk four- to fivefold (Schnack et al. 2010). Orchidopexy does not remove the cancer risk (Myrup et al. 2007), which is understandable, because testis cancer has a fetal origin. Aggregation of cryptorchidism and testis cancer points to a common origin. Familial aggregation and twin studies suggest strongly that the underlying reasons are maternally mediated environmental influences (Schnack et al. 2008; Jensen et al. 2010). Since the testicular descent is hormonally regulated, it is natural to hypothesize that endocrine disruption of the regulatory system can cause cryptorchidism. Animal experiments have revealed an ever-increasing number of chemical compounds that can interfere with testicular descent. We asked, therefore, whether there was any association between exposure to these chemicals and the risk of cryptorchidism. To answer the question we, together with several collaborators, measured levels of more than 100 chemicals in placenta and breast milk samples in the Danish-Finnish joint birth cohort study described under the "Hypospadias" section above.

Finnish and Danish children have specific exposure patterns, as judged by the chemical contents of breast milk (Krysiak-Baltyn et al. 2010). These distinct chemical signatures demonstrate that regional differences in exposures to endocrine disrupters can be significant. Analysis of exposure-cryptorchidism relationships showed weak positive associations with several groups of chemicals, particularly in the Danish population. Polybrominated diphenyl ethers are mainly used as flame retardants. Breast milk concentrations of these chemicals correlated positively with the risk of cryptorchidism (Main et al. 2007). These compounds have anti-androgenic properties, making it biologically plausible that they could contribute to male reproductive problems, although epidemiological association cannot prove any causal relationships. Chlorinated compounds that were used as pesticides in the past and banned in most countries many decades ago continue to contaminate breast milk, because they are so persistent. Although the concentrations have decreased from the peak levels in the 1970s, we still harbor a lot of DDT, DDE and other pesticides in our fat tissue. Concentrations of chlorinated pesticides tended to be higher in breast milk of cryptorchid boys compared with controls, and the Monte Carlo permutation showed a significant association of the sum of eight of the most prevalent pesticides with the risk of cryptorchidism (Damgaard et al. 2006).

Our unpublished results show a similar association of cryptorchidism with dioxin concentrations but not with PCBs. Several phthalate esters have been shown to impair testosterone production and cause reproductive disorders in male rats (Fisher et al. 2003; Foster 2006). These compounds are used mainly as plastisizers and exposure is ubiquitous. However, phthalates are not persistent and their metabolism and clearance are fast. Therefore, it is difficult to estimate exposure accurately and the levels may vary substantially over a short time. In the Danish-Finnish birth cohort study, the mothers collected breast milk samples as small aliquots into glass or porcelain cups after feeding their babies, and they poured the aliquot into a pyrex bottle on top of milk stored and frozen earlier. Thus the milk sample was a longitudinal sample presenting exposure over 1–2 months. Short-term variations in exposures should have smaller effects in this kind of setting. On the other hand, the milk can be considered a good proxy for fetal exposure to persistent compounds that are released from maternal fat during lactation. Phthalate concentrations were not associated with the risk of cryptorchidism, but they showed a positive correlation to LH-testosterone ratio, again suggesting a relative dysfunction of Leydig cells (Main et al. 2006). Swan et al. (2005) reported a negative correlation between maternal urinary phthalate levels and anogenital distance in the male offspring, also suggesting a possible anti-androgenic effect of phthalates in humans. All epidemiological associations between exposures to endocrine disrupters and cryptorchidism are rather weak, and it is unlikely that there is a single culprit among chemicals or compound groups that would cause the disorder. In that sense it is remarkable that any associations with individual chemicals can be seen, and it very possible that combined exposure to hundreds of chemicals may pose a significant risk to children. The incidence of cryptorchidism in Finland and Denmark differs in the same way as hypospadias rate and the incidence of testicular cancer (Fig. 2; Boisen et al. 2004, 2005).

Semen Quality

Occasional reports on declining semen quality started to appear in the 1970s, but they were mostly ignored until Carlsen et al. (1992) reported a meta-analysis of semen studies on normal men published between 1930 and 1990. This analysis showed a dramatic decline in sperm concentrations over the time period, which prompted a lot of new research in this area. The meta-analysis was repeated by others and essentially the same results were published again (Olsen et al. 1995; Swan et al. 1997). Studies from France and Scotland indicated a birth cohort-related decline in sperm concentrations (Auger et al. 1995; Irvine et al. 1996). However, there were also some areas without apparent deterioration of semen quality, e.g., Finland (Suominen and Vierula 1993; Vierula et al. 1996). The study on first pregnancy planners also showed a significant difference in semen quality between Finland and Denmark: Finnish men had higher sperm counts than Danes (Jensen et al. 2000). This study was not performed with external quality controls between the centers, and therefore new initiatives were made to standardize the analytical methods between laboratories. Large regional differences in semen quality were found in these multinational studies (Fig. 3; Jørgensen et al. 2001, 2002). Partners of pregnant women, i.e., fertile men in Finland, had higher sperm counts than those in France or Scotland, and the lowest sperm counts were found in Denmark (Jørgensen et al. 2001). Since the men in this study were selected for fertility, a question remained as to whether men from the general population showed a similar difference. It is difficult to avoid any selection bias in semen studies of general populations, and therefore a group of young men aged 18–20 was selected. At this age the men usually have no idea of their fertility potential and their participation in the study is motivated by curiosity and the small reimbursement fee. Young men from Finland and Estonia showed much better semen quality than those from Denmark and Norway (Jørgensen et al. 2002). The difference reflected the incidence rate of testis cancer in these countries: high in Denmark and Norway and low in

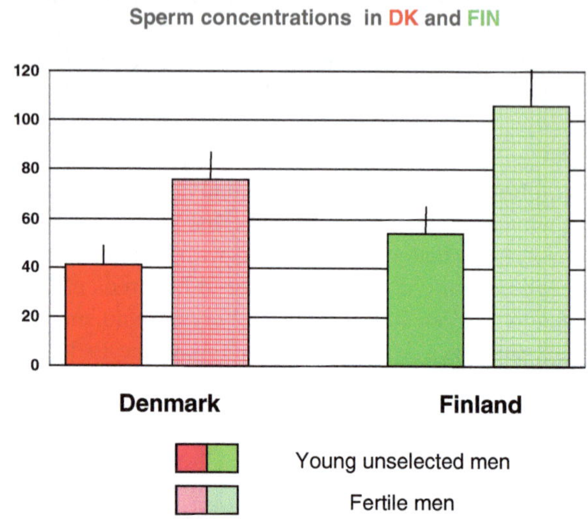

Fig. 3 Semen quality in Denmark and Finland (Jørgensen et al. 2001, 2002)

Finland and Estonia (Adami et al. 1994). While the country differences were clear, it was even more striking how much lower the sperm counts of the young men were compared to older, fertile men. Therefore, longitudinal studies in both the same men and new cohorts were started. The longitudinal study of the same Danish men demonstrated that semen quality did not change much over time, i.e., when the men grew older (Carlsen et al. 2005), so the reason for the difference between the young men and fertile men was not due to their age. Follow-ups of successive Danish cohorts of young men have shown some variation over the years but no clear decline recently (Jørgensen et al. 2006), whereas in Finland semen quality deteriorated significantly during the first decade of this millennium (Jørgensen et al. 2011). At the same time, the incidence rate of testis cancer increased very quickly in Finland (Bray et al. 2006a; Jørgensen et al. 2011). These trends suggest that the difference in male reproductive health between Denmark and Finland may disappear in the near future, and Finland is following the same bad development as other European countries. With a high incidence of testis cancer, Germany shows a similar poor semen quality as Denmark and Norway (Paasch et al. 2008). Semen quality and testis cancer incidence are not invariably correlated. In Japan, testis cancer is very rare, but semen quality is no better than in Europe (Iwamoto et al. 2006). There is no doubt that genetic background is also influencing susceptibility to testis cancer, and this predisposition may differ between semen quality and cancer.

Spermatogenesis can be disturbed by contemporary toxic effects, such as cancer chemotherapy and irradiation, and also by occupational and environmental exposures. Dibromo chloro propane caused severe spermatogenic damage and even sterility in exposed manufacturers (Whorton et al. 1977) and possibly also in farm workers who used it as a nematocide in fruit plantations. These kinds of dramatic adverse effects are not common any longer, but endocrine disrupters may cause subtler effects that are reflected in low sperm counts. Sperm production capacity depends on the number of somatic Sertoli cells in the seminiferous epithelium (Johnson et al. 1984), and these cells cease proliferation by the onset of spermatogenesis. The rapid increase in the Sertoli cell number occurs during early fetal development, in the perinatal period and possibly around the onset of puberty. These periods are therefore very sensitive to external disruption. Mocarelli et al. (2010) reported recently that exposure to high levels of dioxin through breast milk was associated with impaired semen quality, suggesting that testicular development during early postnatal life is very sensitive to endocrine disruption. Fetal exposure to tobacco smoke via mother's smoking has been shown to be associated with impaired semen quality in the son in several studies (Jensen et al. 2004b, 2005; Ravnborg et al. 2011). Thus, there is good evidence about developmental disruption of the testis that results in compromised sperm production capacity. Adult lifestyle factors also contribute; for example, obesity is associated with relative hypogonadism and impaired spermatogenesis (Jensen et al. 2004a). Excessive alcohol consumption (Pajarinen et al. 1997) and drug abuse also damage the testis. However, the lifestyle factors can be corrected and the adverse changes can be reversible, whereas the developmental effects are permanent.

Testicular Dysgenesis Syndrome (TDS)

Epidemiological, biological and experimental evidence indicates an association between testis cancer, hypospadias, cryptorchidism and semen quality (Skakkebæk et al. 2001). While hypospadias and congenital cryptorchidism are developmental disorders by definition, testis cancer and impaired semen quality also have developmental origins. Spermatogenesis can be disturbed by contemporary adverse effects, but on a population level, developmental influences determine spermatogenic capacity. A nationwide study including more than two million Danish men demonstrated familial coaggregation of cryptorchidism, hypospadias and testicular germ cell cancer (Schnack et al. 2010). The aggregation occurred on the individual level, so if a man had had cryptorchidism or hypospadias, his increased relative risk (RR) of developing testis cancer was 3.71 (95% confidence interval (CI) 3.29–4.19) and 2.13 (CI 1.26–3.61), respectively (Schnack et al. 2010). However, the first- and second-degree relatives of the cryptorchidism patients did not have an increased risk for testis cancer, suggesting that environmental effects on the development of the individual contributed to cryptorchidism, hypospadias and testicular germ cell cancer rather than genetic factors. In a detailed study of familial aggregation of cryptorchidism, Schnack et al. (2008) found that recurrence risk ratio (RRR) in same-sex twins was 10.1 (CI 7.78–13.1), in full brothers 3.52 (CI 3.26–3.79), maternal half brothers 2.12 (CI 1.74–2.60), paternal half brothers 1.28 (CI 1.01–1.61), and, in the offspring of a father with a history of cryptorchidism, 2.31 (CI 2.09–2.54). The zygosity of the twins was not analyzed in this study. The data indicated that maternal factors operating in utero are important for the risk of cryptorchidism. Another large Danish study (Jensen et al. 2010), based on more than one million boys and the national twin registry, analyzed the concordance rates of cryptorchidism (i.e., the probability that a pair of individuals will both have cryptorchidism when one of them has it). The concordance rates were the following: boys with no relation 3.2% (CI 2.7–3.6%), paternal half-brothers 3.4% (2.3–4.7%), maternal half-brothers 6.0% (CI 4.5–7.7%), full brothers 8.8% (CI 8.3–9.8%), dizygotic twin brothers 24.1% (CI 16.0–33.6%), and monozygotic twin brothers 27.3% (CI 15.5–41.2%). Birth weight was also considered in this study, and exclusion of low birth weight children did not change the conclusions. The difference between the concordance rates of monozygotic and dizygotic twins is generally used for calculations of the genetic contribution to morbidity. The Danish twin study indicates clearly that environmental factors operating during pregnancy, rather than genetic predisposition, are important for the risk of cryptorchidism. Both population-based studies (Schnack et al. 2010; Jensen et al. 2010) indicate that maternal factors contribute to the risk significantly. Although the twin study did not find a strong genetic component, this does not exclude gene-environment interaction as an important determinant, and in some cases a monogenetic reason can be identified.

Exposure of rats to dibutyl phthalate in utero produces a TDS-like outcome (Fisher et al. 2003), although human-like testis cancer does not develop in rodent models. The same adverse effects can be seen in all animal experiments where the dams are exposed to anti-androgenic substances; furthermore, the compounds act in a dose-additive manner, rendering even low doses of components of the exposure mixtures harmful (Christiansen et al. 2008; Rider et al. 2008, 2009). Normal hormonal control is essential for proper development and function of genital organs. Therefore, it is easy to understand that disruption of this endocrine system can result in several adverse outcomes, each of which may have a somewhat different susceptibility time window. Thus, cryptorchidism and hypospadias may differ in sensitive developmental time periods and spermatogenic capacity is determined during a large time window spanning from fetal development all the way to the onset of puberty. Realization of these time frames is important for the planning of studies and also for counselling the public about possible ways to prevent reproductive health problems.

Conclusions and Future Perspectives

Current male reproductive health is poor and infertility will be an increasing problem. The reasons are not known, but emerging evidence from experimental animal studies, wildlife, and epidemiological studies combined with our biological and clinical understanding support the hypothesis that environmental endocrine disrupters contribute to the problem. It is therefore important to increase awareness of male reproductive health problems and strengthen interdisciplinary, translational research in this field, as specified recently by the European Science Foundation

(ESF; ESF Science Policy Briefing, September 2010, www.esf.org). Long-term epidemiological studies combining genetic and environmental information with good clinical examinations, biological pathway analyses and effective bioinformatics will be crucial to understanding the etiology of poor male reproductive health. Targeted research efforts to better understand the mechanisms involved in these processes, including late effects of fetal exposure, will provide paths forward for better male reproductive health.

Acknowledgments We thank Dr. Niels Jørgensen for making Fig. 1 available to us. The studies were supported by the Academy of Finland, the Sigrid Jusélius Foundation, EU the European Commission (QLK4-CT-1999-01422, QLK4-CT-2001-8 00269, QLK4-2002-00603, FP7/2008-2012: DEER 212844), the Danish Medical Research Council (9700833, 9700909), the Svend Andersen's, Velux and Novo Nordisk Foundations, and the Danish Network on Endocrine Disrupters (DAN-ED).

References

Adami H-O, Bergström R, Möhner M, Zatonski W, Storm H, Ekbom A, Tretli S, Teppo L, Ziegler H, Rahu M, Gurevicius R, Stengrevics A (1994) Testicular cancer in nine Northern European countries. Int J Cancer 59:33–38

Aho M, Koivisto AM, Tammela TL, Auvinen A (2000) Is the incidence of hypospadias increasing? Analysis of Finnish hospital discharge data 1970–1994. Environ Health Perspect 108:463–465

Auger J, Kunstmann JM, Czyglik F, Jouannet P (1995) Decline in semen quality among fertile men in Paris during the past 20 years. N Engl J Med 332:281–285

Bergström R, Adami H-O, Möhner M, Zatonski W, Storm H, Ekbom A, Tretli S, Teppo L, Akre O, Hakulinen T (1996) Increase in testicular cancer incidence in six European countries: a birth cohort phenomenon. J Natl Cancer Inst 88:727–733

Boisen KA, Kaleva M, Main KM, Virtanen HE, Haavisto A-M, Schmidt IM, Chellakooty M, Damgaard IN, Mau C, Reunanen M, Skakkebæk NE, Toppari J (2004) Difference in prevalence of congenital cryptorchidism in infants between two Nordic countries. Lancet 363:1264–1269

Boisen KA, Chellakooty M, Schmidt IM, Kai CM, Damgaard IN, Suomi AM, Toppari J, Skakkebæk NE, Main KM (2005) Hypospadias in a cohort of 1072 Danish newborn boys: prevalence and relationship to placental weight, anthropometrical measurements at birth, and reproductive hormone levels at 3 months of age. J Clin Endocrinol Metab 90:4041–4046

Bray F, Ferlay J, Devesa SS, McGlynn KA, Møller H (2006a) Interpreting the international trends in testicular seminoma and nonseminoma incidence. Natl Clin Pract Urol 3:532–543

Bray F, Richiardi L, Ekbom A, Pukkala E, Cuninkova M, Moller H (2006b) Trends in testicular cancer incidence and mortality in 22 European countries: continuing increases in incidence and declines in mortality. Int J Cancer 118:3099–3111

Carlsen E, Giwercman A, Keiding N, Skakkebæk NE (1992) Evidence for decreasing quality of semen during past 50 years. BMJ 305:609–613

Carlsen E, Swan SH, Holm PJ, Skakkebæk NE (2005) Longitudinal changes in semen parameters in young Danish men from the Copenhagen area. Hum Reprod 20:942–949

Christiansen S, Scholze M, Axelstad M, Boberg J, Kortenkamp A, Hass U (2008) Combined exposure to anti-androgens causes markedly increased frequencies of hypospadias in the rat. Int J Androl 31:241–248

Damgaard IN, Skakkebæk NE, Toppari J, Virtanen HE, Shen H, Schramm KW, Petersen JH, Jensen TK, The Nordic Cryptorchidism Study Group, Main KM (2006) Persistent pesticides in human breast milk and cryptorchidism. Environ Health Perspect 114:1133–1138

Fisher JS, Macpherson S, Marchetti N, Sharpe RM (2003) Human "testicular dysgenesis syndrome": a possible model using in-utero exposure of the rat to dibutyl phthalate. Hum Reprod 18:1383–1394

Foster PM (2006) Disruption of reproductive development in male rat offspring following in utero exposure to phthalate esters. Int J Androl 29:140–147

Goriely A, Hansen RM, Taylor IB, Olesen IA, Jacobsen GK, McGowan SJ, Pfeifer SP, McVean GA, Rajpert-De Meyts E, Wilkie AO (2009) Activating mutations in FGFR3 and HRAS reveal a shared genetic origin for congenital disorders and testicular tumors. Nat Genet 41:1247–1252

Hadziselimovic F, Hocht B, Herzog B, Buser MW (2007) Infertility in cryptorchidism is linked to the stage of germ cell development at orchidopexy. Horm Res 68:46–52

Hardell L, van Bavel B, Lindstrom G, Carlberg M, Dreifaldt AC, Wijkstrom H, Starkhammar H, Eriksson M, Hallquist A, Kolmert T (2003) Increased concentrations of polychlorinated biphenyls, hexachlorobenzene, and chlordanes in mothers of men with testicular cancer. Environ Health Perspect 111:930–934

Hardell L, van Bavel B, Lindstrom G, Eriksson M, Carlberg M (2006) In utero exposure to persistent organic pollutants in relation to testicular cancer risk. Int J Androl 29:228–234

Hass U, Scholze M, Christiansen S, Dalgaard M, Vinggaard AM, Axelstad M, Metzdorff SB, Kortenkamp A (2007) Combined exposure to anti-androgens exacerbates disruption of sexual differentiation in the rat. Environ Health Perspect 115(Suppl 1):122–128

Irvine S, Cawood E, Richardson D, MacDonald E, Aitken J (1996) Evidence of deteriorating semen quality in the United Kingdom: birth cohort study in 577 men in Scotland over 11 years. Br Med J 312:467–471

Iwamoto T, Hoshino T, Nishida T, Baaba K, Matsusita T, Kaneko S, Tanaka N, Naka M, Yoshiike M, Nozawa S, Skakkebæk NE, Jørgensen N (2006) Semen quality of 324 fertile Japanese men. Hum Reprod 21:760–765

Jensen TK, Vierula M, Hjollund NHI, Saaranen M, Scheike T, Saarikoski S, Suominen J, Keiski A, Toppari J, Skakkebæk NE, The Danish First Pregnancy Planner Study Team (2000) Semen quality among Danish and Finnish men attempting to conceive. Eur J Endocrinol 142:47–52

Jensen TK, Andersson A-M, Jørgensen N, Andersen A-G, Carlsen E, Petersen JH, Skakkebæk NE (2004a) Body mass index in relation to semen quality and reproductive hormones among 1,558 Danish men. Fertil Steril 82:863–870

Jensen TK, Jørgensen N, Punab M, Haugen TB, Suominen J, Zilaitiene B, Horte A, Andersen A-G, Carlsen E, Magnus Ø, Matulevicius V, Nermoen I, Vierula M, Keiding N, Toppari J, Skakkebæk NE (2004b) Association of in utero exposure to maternal smoking with reduced semen quality and testis size in adulthood: a cross-sectional study of 1,770 young men from the general population in five European countries. Am J Epidemiol 159:49–58

Jensen MS, Mabeck LM, Toft G, Thulstrup AM, Bonde JP (2005) Lower sperm counts following prenatal tobacco exposure. Hum Reprod 20:2559–2566

Jensen MS, Toft G, Thulstrup AM, Henriksen TB, Olsen J, Christensen K, Bonde JP (2010) Cryptorchidism concordance in monozygotic and dizygotic twin brothers, full brothers, and half-brothers. Fertil Steril 93:124–129

Johnson L, Zane RS, Petty CS, Neaves WB (1984) Quantification of the human Sertoli cell population: its distribution, relation to germ cell numbers, and age-related decline. Biol Reprod 31:785–795

Jørgensen N, Andersen A-G, Eustache F, Irvine DS, Suominen J, Petersen JH, Andersen AN, Auger J, Cawood EHH, Horte A, Jensen TK, Jouannet P, Keiding N, Vierula M, Toppari J, Skakkebæk NE (2001) Regional differences in semen quality in Europe. Hum Reprod 16:1012–1019

Jørgensen N, Carlsen E, Nermoen I, Punab M, Suominen J, Andersen A-G, Andersson A-M, Haugen TB, Horte A, Jensen TK, Magnus Ø, Petersen JH, Vierula M, Toppari J, Skakkebæk NE (2002)

East-West gradient in semen quality in the Nordic-Baltic area: a study of men from the general population in Denmark, Norway, Estonia and Finland. Hum Reprod 17:2199–2208
Jørgensen N, Asklund C, Carlsen E, Skakkebæk NE (2006) Coordinated European investigations of semen quality: results from studies of Scandinavian young men is a matter of concern. Int J Androl 29:54–61
Jørgensen N, Vierula M, Jacobsen R, Pukkala E, Perheentupa A, Virtanen HE, Skakkebæk NE, Toppari J (2011) Recent adverse trends in semen quality and testis cancer incidence among Finnish men. Int J Androl 34:e37–e48. doi:10.1111/j.1365-2605.2010.01133.x
Kanetsky PA, Mitra N, Vardhanabhuti S, Li M, Vaughn DJ, Letrero R, Ciosek SL, Doody DR, Smith LM, Weaver J, Albano A, Chen C, Starr JR, Rader DJ, Godwin AK, Reilly MP, Hakonarson H, Schwartz SM, Nathanson KL (2009) Common variation in KITLG and at 5q31.3 predisposes to testicular germ cell cancer. Nat Genet 41:811–815
Kollin C, Karpe B, Hesser U, Granholm T, Ritzen EM (2007) Surgical treatment of unilaterally undescended testes: testicular growth after randomization to orchiopexy at age 9 months or 3 years. J Urol 178:1589–1593
Krysiak-Baltyn K, Toppari J, Skakkebæk NE, Jensen TS, Virtanen HE, Schramm KW, Shen H, Vartiainen T, Kiviranta H, Taboureau O, Brunak S, Main KM (2010) Country-specific chemical signatures of persistent environmental compounds in breast milk. Int J Androl 33:270–278
Lee PA (1993) Fertility in chryptorchidism. Endocrinol Metab Clin North Am 33(3):479–490
Lee PA (2005) Fertility after cryptorchidism: epidemiology and other outcome studies. Urology 66:427–431
Llanes GL, Lujan GM, Rodriguez GN, Garcia TA, Berenguer SA (2008) Trends in the incidence of testicular germ cell cancer in a 300.000 inhabitants Spanish population (1991–2005). Actas Urol Esp 32:691–695
Lund L, Engebjerg MC, Pedersen L, Ehrenstein V, Norgaard M, Sorensen HT (2009) Prevalence of hypospadias in Danish boys: a longitudinal study, 1977–2005. Eur Urol 55:1022–2026
Main KM, Mortensen GK, Kaleva M, Boisen K, Damgaard I, Chellakooty M, Schmidt IM, Suomi A-M, Virtanen H, Petersen JH, Andersson A-M, Toppari J, Skakkebæk NE (2006) Human breast milk contamination with phthalates and alterations of endogenous reproductive hormones in three months old infants. Environ Health Perspect 114:270–276
Main KM, Kiviranta H, Virtanen HE, Sundqvist E, Tuomisto JT, Tuomisto J, Vartiainen T, Skakkebæk NE, Toppari J (2007) Flame retardants in placenta and breast milk and cryptorchidism in Newborn boys. Environ Health Perspect 115:1519–1526
Mocarelli P, Gerthoux PM, Needham LL, Patterson DG Jr, Limonta G, Falbo R, Signorini S, Bertona M, Crespi C, Sarto C, Scott PK, Turner WE, Brambilla P (2010) Perinatal exposure to low doses of dioxin can permanently impair human semen quality. Environ Health Perspect. doi:10.1289/ehp. 1002134
Myrup C, Schnack TH, Wohlfahrt J (2007) Correction of cryptorchidism and testicular cancer. N Engl J Med 357:825–827
Olsen GW, Bodner KM, Ramlow JM, Ross CE, Lipshultz LI (1995) Have sperm counts been reduced 50 percent in 50 years? A statistical model revisited. Fertil Steril 63:887–893
Paasch U, Salzbrunn A, Glander HJ, Salzbrunn H, Grunewald S, Stucke J, Skakkebæk NE, Jørgensen N (2008) Semen quality in sub-fertile range for a significant proportion of young men from the general German population: a co-ordinated, controlled study of 791 men from Hamburg and Leipzig. Int J Androl 31:93–102
Pajarinen J, Laippala P, Penttila A, Karhunen PJ (1997) Incidence of disorders of spermatogenesis in middle aged Finnish men, 1981–91: two necropsy series. Br Med J 314:13–18
Pierik FH, Burdorf A, Nijman JM, de Muinck Keizer-Schrama SM, Juttmann RE, Weber RF (2002) A high hypospadias rate in the Netherlands. Hum Reprod 17:1112–1115
Quigley CA, DeBellis A, Marschke KB, El-Awady MK, Wilson EM, French FS (1995) Androgen receptor defects: historical, clinical, and molecular perspectives. Endocr Rev 16:271–321

Rajpert-De Meyts E (2006) Developmental model for the pathogenesis of testicular carcinoma in situ: genetic and environmental aspects. Hum Reprod Update 12:303–323

Rapley EA, Turnbull C, Al Olama AA, Dermitzakis ET, Linger R, Huddart RA, Renwick A, Hughes D, Hines S, Seal S, Morrison J, Nsengimana J, Deloukas P, Rahman N, Bishop DT, Easton DF, Stratton MR (2009) A genome-wide association study of testicular germ cell tumor. Nat Genet 41:807–810

Ravnborg TL, Jensen TK, Andersson AM, Toppari J, Skakkebaek NE, Jorgensen N (2011) Prenatal and adult exposures to smoking are associated with adverse effects on reproductive hormones, semen quality, final height and body mass index. Hum Reprod. doi:10.1093/humrep/der011

Rey RA, Codner E, Iniguez G, Bedecarras P, Trigo R, Okuma C, Gottlieb S, Bergada I, Campo SM, Cassorla FG (2005) Low risk of impaired testicular sertoli and leydig cell functions in boys with isolated hypospadias. J Clin Endocrinol Metab 90:6035–6040

Rider CV, Furr J, Wilson VS, Gray LE Jr (2008) A mixture of seven antiandrogens induces reproductive malformations in rats. Int J Androl 31:249–262

Rider CV, Wilson VS, Howdeshell KL, Hotchkiss AK, Furr JR, Lambright CR, Gray LE Jr (2009) Cumulative effects of in utero administration of mixtures of antiandrogens on male rat reproductive development. Toxicol Pathol 37:100–113

Schnack TH, Zdravkovic S, Myrup C, Westergaard T, Wohlfahrt J, Melbye M (2008) Familial aggregation of cryptorchidism–a nationwide cohort study. Am J Epidemiol 167:1453–1457

Schnack TH, Poulsen G, Myrup C, Wohlfahrt J, Melbye M (2010) Familial coaggregation of cryptorchidism, hypospadias, and testicular germ cell cancer: a nationwide cohort study. J Natl Cancer Inst 102:187–192

Skakkebæk NE, Rajpert-De Meyts E, Main KM (2001) Testicular dysgenesis syndrome: an increasingly common developmental disorder with environmental aspects. Hum Reprod 16:972–978

Skakkebæk NE, Jørgensen N, Main KM, Rajpert-De Meyts E, Leffers H, Andersson A-M, Juul A, Carlsen E, Mortensen GK, Jensen TK, Toppari J (2006) Is human fecundity declining? Int J Androl 29:2–11

Suomi A-M, Main KM, Kaleva M, Schmidt IM, Chellakooty M, Virtanen HE, Boisen KA, Damgaard IN, Kai CM, Skakkebæk NE, Toppari J (2006) Hormonal changes in 3-month-old cryptorchid boys. J Clin Endocrinol Metab 91:953–958

Suominen J, Vierula M (1993) Semen quality of Finnish men. Br Med J 306:1579

Swan SH, Elkin EP, Fenster L (1997) Have sperm densities declined? A reanalysis of global trend data. Environ Health Perspect 105:1228–1232

Swan SH, Main KM, Liu F, Stewart SL, Kruse RL, Calafat AM, Mao CS, Redmon JB, Ternand CL, Sullivan S, Teague JL (2005) Decrease in anogenital distance among male infants with prenatal phthalate exposure. Environ Health Perspect 113:1056–1061

Taskinen S, Hovatta O, Wikstrom S (1997) Sexual development in patients treated for cryptorchidism. Scand J Urol Nephrol 31:361–364

Toppari J, Kaleva M, Virtanen HE (2001) Trends in the incidence of cryptorchidism and hypospadias, and methodological limitations of registry-based data. Hum Reprod Update 7:282–286

van der Zanden LF, van Rooij I, Feitz WF, Knight J, Donders AR, Renkema KY, Bongers EM, Vermeulen SH, Kiemeney LA, Veltman JA, Arias-Vasquez A, Zhang X, Markljung E, Qiao L, Baskin LS, Nordenskjold A, Roeleveld N, Franke B, Knoers NV (2011a) Corrigendum: common variants in DGKK are strongly associated with risk of hypospadias. Nat Genet 43:277

van der Zanden LF, van Rooij I, Feitz WF, Knight J, Donders AR, Renkema KY, Bongers EM, Vermeulen SH, Kiemeney LA, Veltman JA, Arias-Vasquez A, Zhang X, Markljung E, Qiao L, Baskin LS, Nordenskjold A, Roeleveld N, Franke B, Knoers NV (2011b) Common variants in DGKK are strongly associated with risk of hypospadias. Nat Genet 43:48–50

Vierula M, Niemi M, Keiski A, Saaranen M, Saarikoski S, Suominen J (1996) High and unchanged sperm counts of Finnish men. Int J Androl 19:11–17

Virtanen HE, Kaleva M, Haavisto A-M, Schmidt IM, Chellakooty M, Main KM, Skakkebæk NE, Toppari J (2001) The birth rate of hypospadias in the Turku area in Finland. APMIS 109:96–100

Whorton D, Krauss RM, Marshall S, Milby TH (1977) Infertility in male pesticide workers. Lancet 2:1259–1261

Origin of Testicular Dysgenesis Syndrome Disorders in the Masculinization Programming Window: Relevance to Final Testis Size (=Sperm Production)

Richard M. Sharpe, Sarah Auharek, Hayley M. Scott, Luiz Renato de Franca, Amanda J. Drake, and Sander van den Driesche

Abstract Testicular dysgenesis syndrome (TDS) disorders are common and/or increasing in incidence in human males, implicating lifestyle/environmental causes. Our studies in rats suggest that TDS disorders originate in fetal life because of deficient testosterone production by the fetal testis within a specific masculinization programming window (MPW). Administration of dibutyl phthalate (DBP) to pregnant rats suppresses testosterone levels in the fetal testis more dramatically after the MPW than during the MPW, but only suppression in the MPW affects anogenital distance (AGD), confirming that this simple measure provides a lifelong readout of androgen action just within the MPW. Using maternal treatments (DBP, linuron, prochloraz, alone or in combination) that can impair testosterone production by the fetal testis, we show that suppression of androgen action within the MPW, as inferred from AGD, is highly correlated ($P < 0.0001$) with testis size at e21.5 (when Sertoli cells are still proliferating), at postnatal day 25 (early puberty, when final Sertoli cell number has been determined) and in adulthood (when it equates to the level of sperm production). As a similar correlation was found between AGD and final Sertoli cell number, effects on this parameter probably explain the AGD:testis size correlation. Low sperm count is the commonest TDS disorder, affecting ~20% of young men. Emerging evidence suggests it may originate in fetal life in the MPW, consistent with the present rat studies.

R.M. Sharpe (✉)
MRC/University of Edinburgh Centre for Reproductive Health, The Queen's Medical Research Institute, 47 Little France Crescent, Edinburgh EH16 4TJ, UK
e-mail: r.sharpe@ed.ac.uk

J.-P. Bourguignon et al. (eds.), *Multi-System Endocrine Disruption*,
Research and Perspectives in Endocrine Interactions,
DOI 10.1007/978-3-642-22775-2_11, © Springer-Verlag Berlin Heidelberg 2011

Introduction

Male reproductive disorders that manifest at either birth (incomplete testicular descent, termed cryptorchidism, or malpositioning of the urethral opening on the penis, termed hypospadias) or in young adulthood (low sperm counts, testicular germ cell cancer) are common and/or are increasing in incidence in most Western populations (Skakkebaek et al. 2001, 2007). For various reasons, such as shared risk factors and being risk factors for each other, these disorders have been hypothesized to comprise a testicular dysgenesis syndrome (TDS) with a common fetal origin that may involve deficient testosterone (androgen) production by the fetal testis (Skakkebaek et al. 2001; Sharpe and Skakkebaek 2008). This hypothesis has been strongly supported by animal experimental studies which, for example, show that fetal exposure of rats to certain phthalate esters (a group of ubiquitous environmental chemicals) can induce a TDS-like spectrum of disorders in the male offspring, with increased rates of cryptorchidism, hypospadias and reduced testis size (which equates to reduced sperm production; Fisher et al. 2003; Foster 2006; Gray et al. 2006). These changes are associated with reduced production of testosterone by the fetal testis. Therefore, a key aim in research into the origins of male reproductive disorders is to identify when in fetal life such disorders may arise, the importance of androgen action at this time (as opposed to at any other times in life) and how and when it can be perturbed and with what long-term consequences.

Identification and Key Importance of the Masculinization Programming Window

Arguably the most important step forward in the TDS concept has been the demonstration that normal development of the male reproductive system is "programmed" by androgen action within an early fetal time window termed the masculinization programming window (MPW; Welsh et al. 2008). Deficient androgen production or action within the MPW (~e15.5–e17.5 in rats), but not before or after, increases the incidence of cryptorchidism and hypospadias in males and reduces testis size in adulthood; it also impacts the ultimate size of other male reproductive tract organs (penis, prostate, seminal vesicles) (Fig. 1; Welsh et al. 2008, 2010; Drake et al. 2009; MacLeod et al. 2010). These programming effects appear to be irreversible, that is, they cannot be rescued by exogenous androgen treatment after the MPW (Welsh et al. 2010; MacLeod et al. 2010). Another key endpoint that appears to be programmed by androgen action in the MPW is the male-specific increase in anogenital distance (AGD; Fig. 1), which is ~twice as long in normal adult males as in females in both rodents and humans. The potential importance of AGD is that it provides a (non-invasive) life-long readout of androgen exposure during the MPW, once there is due allowance for age and bodyweight. This is reinforced by the demonstration in experimental studies in rats that the risk

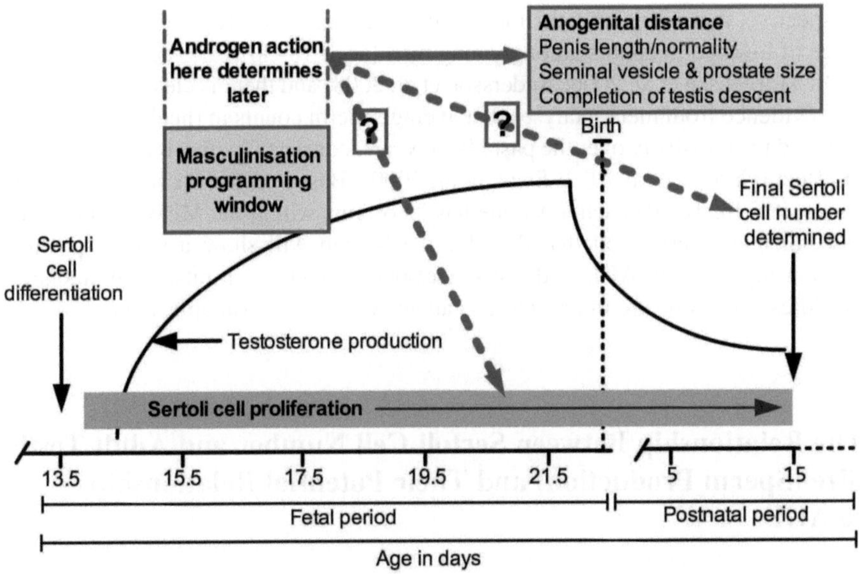

Fig. 1 Schematic diagram to illustrate the relationships between fetal/postnatal age, testosterone production by the fetal/neonatal testis, the masculinization programming window (MPW) and Sertoli cell differentiation and subsequent proliferation in the rat. The *dashed lines* emanating from the MPW indicate effects that are considered likely but which are unproven. Other details are provided in the text

and severity of cryptorchidism and hypospadias are both correlated inversely with AGD, whereas penis, seminal vesicle, prostate and testis size are all positively correlated with AGD at all postnatal ages from prepuberty through to adulthood (Welsh et al. 2008, 2010; Drake et al. 2009; MacLeod et al. 2010). Emerging evidence in humans shows a similar relationship between AGD and reproductive organ size and normality (Swan et al. 2005; Hsieh et al. 2008; Thankamony et al. 2009); the MPW in humans is thought to be within the period of 8–14 weeks of gestation, which is the end of the first trimester + the start of the second trimester (Welsh et al. 2008).

What the findings discussed above collectively demonstrate is that androgen action within the MPW exerts a lifelong constraint on reproductive organ development and its ultimate size (Fig. 1). The implication is that any deficiency in androgen production or action within this critical time window will have irreparable reproductive consequences. This finding fits closely with the TDS hypothesis and probably verifies it (Sharpe and Skakkebaek 2008). Nevertheless, several fundamentally important questions remain for which we currently have no answers. For example, what determines the timing of the MPW, and what pathways are triggered specifically within the MPW by androgens that cannot be triggered before or after? Another puzzle is how androgen action within the MPW can determine ultimate testis size and thus the ceiling of sperm production. From the human health perspective, this is arguably the most important puzzle to solve because the prevalence of low sperm counts among young men (aged 18–30 years) across Europe is ~20% (Jørgensen et al. 2006; Andersson et al. 2008) and there is clear, if controversial, evidence from meta-analyses that average sperm counts in (healthy) men have declined progressively over the past 50–60 years, consistent with a birth cohort type of effect (Carlsen et al 1992; Swan et al. 2000). Based on the rat studies outlined above, it is likely that such a decline has its origins within the MPW. If this is the case, then it is predicted that data for the human will show a similar positive relationship between AGD and testis size/sperm counts as in rats. Several studies to address this issue are in progress, so an answer will be available soon.

The Relationship Between Sertoli Cell Number and Adult Testis Size (Sperm Production) and Their Potential Relationship to AGD

Both animal and human studies have established that the major determinant of adult testis size and sperm production is the number of Sertoli cells (Sharpe 1994, 2003). These are the cells that, in adulthood, physically and metabolically support the germ cells throughout their development into sperm, and each Sertoli cell can support only a fixed number of germ cells. Therefore, the more Sertoli cells that are present in the testis, the more germ cells can be supported, and therefore the higher will be the level of sperm production. In humans, this relationship will also relate directly to sperm counts in the ejaculate, as humans, in contrast to most animals, do not store sperm and sperm count in the ejaculate is a reflection of what is being produced (Sharpe 1994; Sharpe et al. 2003). Furthermore, as the bulk of the adult testis is made up of germ cells, Sertoli cell number also indirectly determines adult testis size (Sharpe 1994). Therefore, an obvious explanation for the correlation between AGD and final testis size would be if androgen action within the MPW somehow programmed final Sertoli cell number (Fig. 1). This is an attractive concept because Sertoli cell differentiation is what initiates testis formation in fetal life, and these cells are actively proliferating during the MPW. However, it is equally clear that Sertoli cells continue to proliferate long after the MPW, including during the postnatal and prepubertal periods (Fig. 1), and there is good evidence that androgen action is at least partly responsible for Sertoli cell proliferation during these periods (Sharpe et al. 2003; Auharek et al. 2010). Furthermore, in a preliminary study, we did not find any significant relationship between AGD and Sertoli cell number in rats, although this finding was based on small numbers of animals (Scott et al. 2008).

Therefore, although the notion that androgen action in the MPW can shape future sperm counts/production is attractive, and would be wholly consistent with the TDS hypothesis, it is unclear how it could operate mechanistically (Scott et al. 2009; MacLeod et al. 2010). To address these doubts, we undertook two proof-of-concept studies: first, to further validate AGD as a read-out of androgen action only within the MPW, and second, to demonstrate categorically if perturbation of testicular androgen production within the MPW, using three different chemicals alone or together, was reflected in reduced testis size at various ages prenatally and postnatally. Furthermore, we sought to determine whether AGD was correlated with final Sertoli cell number.

Proof-of-Concept Studies

1. Only suppression of testosterone production by the fetal testis during the MPW affects AGD

   AGD has been used for many years by toxicologists as an index of overall androgen exposure in fetal life, whereas what our studies propose is that this time frame is not accurate and that it reflects only deficient androgen exposure within the MPW. To prove this hypothesis, we treated pregnant rats with a dose of DBP (500 mg/kg/day in corn oil by oral gavage), which is known to suppress testosterone production by the fetal testis, but we varied the treatment period to include just the MPW (e15.5–e17.5), a period after the MPW (e19.5–e20.5) when testosterone levels are normally highest, or for the whole fetal period (e13.5–e20.5). Fetuses were sampled on e21.5 (Fig. 2a, c, d) or, in one instance, on e17.5 after treatment for the previous 2 days with DBP or vehicle (Fig. 2b). This study showed that only DBP treatments that included the period of the MPW resulted in reduced AGD at e21.5 (Fig. 2c). Restricting DBP treatment just to the MPW resulted in a reduction in testosterone levels at e17.5 (Fig. 2b) but not at e21.5 (Fig. 2a). In contrast, DBP treatment just in late gestation (e19.5–e20.5), in a time period after the MPW and in which the male-female difference in AGD first begins to emerge (Welsh et al. 2008), had no effect on AGD (Fig. 2c). All treatment modalities reduced testis weight at e21.5 (Fig. 2d), and earlier studies showed that this reduction was associated with reduced Sertoli cell number (Scott et al. 2008). Therefore, the contrast in effect on AGD of DBP treatment from e15.5–e17.5 (the MPW) versus that from e19.5–e20.5 (after the MPW) shows that reduction in AGD at e21.5 depends on there being suppression of testosterone levels during the MPW, and gross suppression of testosterone levels at e21.5 after the MPW has no impact on AGD (Fig. 2c).
2. Induction of fetal testis deficiency in testosterone production within the MPW, by any means, determines testis size in the rat, probably via effects on Sertoli cell number

   One of the limitations of our experimental studies in animals directed at understanding the relationship between androgen action in the MPW and final testis size (Fig 1; Drake et al. 2009) is that only one experimental treatment modality has been used, namely fetal exposure to dibutyl phthalate (DBP). Although this treatment induces a reduction in testosterone production within

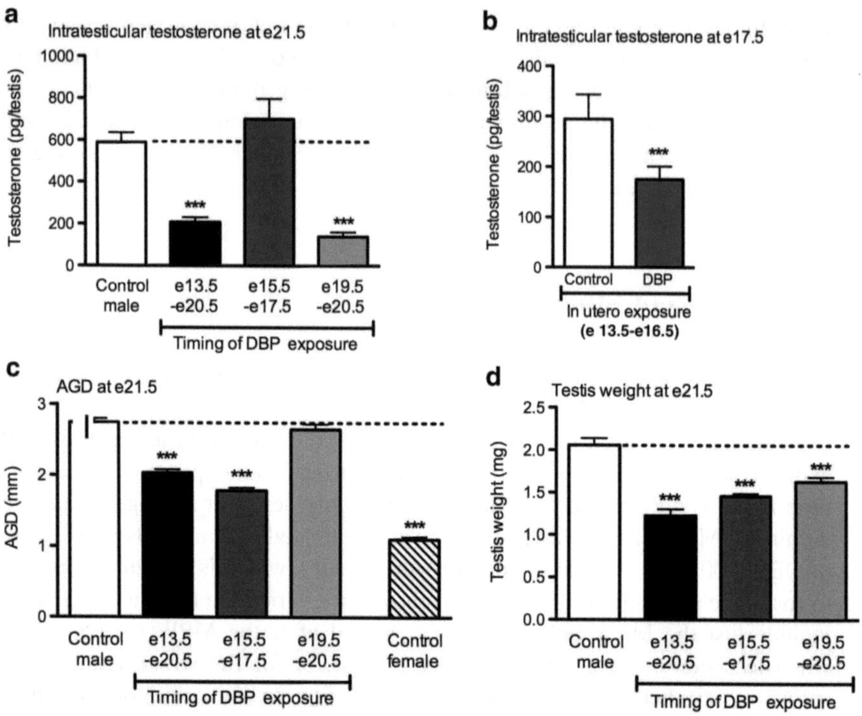

Fig. 2 Effect of maternal treatment of rats with vehicle (control) or DBP (500 mg/kg/day) in different fetal time widows on intratesticular testosterone (ITT) levels (**a**), AGD (**c**) and testis weight (**d**) at e21.5; panel **b** shows that treatment with DBP also suppresses ITT during the MPW. Values are means ± SEM for 23–56 animals per treatment group with the exception of data for ITT at e21.5 in animals exposed to DBP from e15.5–e17.5, when N = 9. ***P < 0.001, in comparison with respective control group (analysis by one way analysis of variance followed by the Bonferroni post-hoc test)

the MPW and a consequent reduction in AGD (Fig. 2) and in Sertoli cell proliferation (Scott et al. 2007, 2008), these effects are often quite small. However, a more important reservation is that DBP may also directly target Sertoli cells (Foster 2006), so that its effect on Sertoli cell proliferation might occur, wholly or partly, independently of its suppression of testosterone production during the MPW. Therefore, to circumvent this problem, we have undertaken a series of studies in which pregnant female rats have been treated with one or more chemicals (prochloraz, linuron, dibutyl phthalate) that have been shown to suppress testosterone production by the fetal testis (Blystone et al. 2007; Scott et al. 2007; Wilson et al. 2009) or with the vehicle (corn oil) as a control (Table 1). Treatments were administered during fetal life from the moment of testis differentiation (e13.5–e20.5) ± continued treatment (of the mother) for the first 15 days after birth; treatment from e13.5-postnatal day 15 would encompass the complete period of Sertoli cell proliferation (Fig. 1). For these studies, the androgen receptor antagonist, flutamide, was not used for

Table 1 Maternal treatment modalities used in the present studies to decrease testosterone production by the fetal and/or neonatal testis

Gestational treatment (e13.5–e21.5)[a]	Postnatal treatment (days 1–15)
Corn oil (= control)	Corn oil (= control)
DBP 500 mg/kg/day	Corn oil
DBP 500 mg/kg/day	DBP 500 mg/kg/day
DBP 500 mg/kg/day	Flutamide 100 mg/kg/day
DBP 100 mg/kg/day	Corn oil
DBP 100 mg/kg/day	DBP 100 mg/kg/day
Linuron 50 mg/kg/day	Corn oil
Prochloraz 50 mg/kg/day	Corn oil
Prochloraz 100 mg/kg/day	Corn oil
Linuron 50 mg/kg/day + Prochloraz 50 mg/kg/day	Corn oil
Linuron 50 mg/kg/day + Prochloraz 50 mg/kg/day	Linuron 50 mg/kg/day + Prochloraz 50 mg/kg/day
Linuron 50 mg/kg/day + Prochloraz 50 mg/kg/day + DBP 100 mg/kg/day	Corn oil
Linuron 50 mg/kg/day + Prochloraz 100 mg/kg/day + DBP 100 mg/kg/day	Corn oil
Linuron 50 mg/kg/day + Prochloraz 50 mg/kg/day + DBP 100 mg/kg/day	Linuron 50 mg/kg/day + Prochloraz 50 mg/kg/day + DBP 100 mg/kg/day
Linuron 50 mg/kg/day + Prochloraz 100 mg/kg/day + DBP 100 mg/kg/day	Linuron 50 mg/kg/day + Prochloraz 100 mg/kg/day + DBP 100 mg/kg/day

All chemicals were sourced from Sigma except for corn oil, which was purchased from a supermarket.

[a]For animals sampled on e21.5, maternal treatment stopped on e20.5

treatment during pregnancy, as our studies have shown that it is unable to antagonize androgen action within the fetal testis and it does not itself affect the production of testicular testosterone (Scott et al. 2007). However, it has been shown to antagonize androgen action within the early postnatal testis when administered to the mother (and transferred to the male pups via breast milk), an effect that is probably due to the much lower intratesticular testosterone levels postnatally versus prenatally (Scott et al. 2007; Auharek et al. 2010). Therefore, we also included flutamide as a potential postnatal treatment; in this regard, our studies show that this (and other) postnatal treatments have no effect on AGD (van den Driesche et al. 2011 and unpublished data), as would be predicted.

In these studies, we sampled exposed animals at three different ages: either at e21.5, when Sertoli cells are still actively proliferating; at postnatal day 25, which is ~10 days after final Sertoli cell number has been determined (Fig. 1); or in adulthood, when final testis size (which equates to sperm production) has been determined. At all ages, AGD was used as a read-out of fetal androgen exposure during the MPW (Fig. 2).

Judging by AGD, all of the chemical treatments listed in Table 1 caused reductions in testosterone levels/action within the MPW, although these effects were not large except in animals exposed to the mixture of DBP/linuron/prochloraz

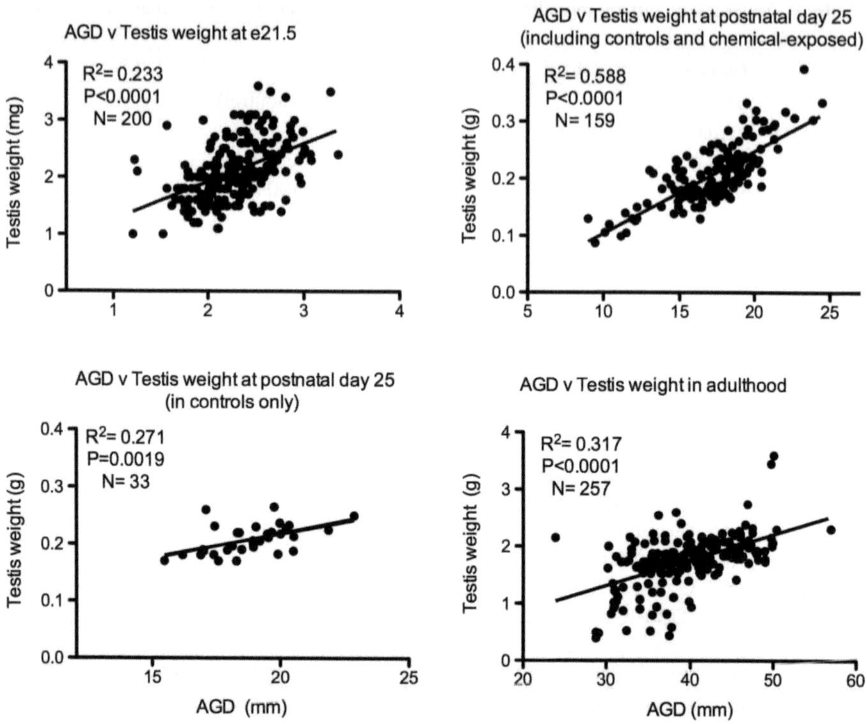

Fig. 3 Relationship between anogenital distance (AGD) and testis weight at e21.5 (*top left*), day 25 (*top right and bottom left*) and in adulthood (*bottom right*) in male offspring from mothers subjected to one of the treatment modalities listed in Table 1. Correlations were determined using linear regression analysis

(data not shown). Nevertheless, these effects were sufficient to affect testis weight and thus allow detailed examination of the relationship between AGD and testis size across a wider numerical range than would be possible in just controls. The results showed that, at all ages studied, AGD was significantly correlated ($P < 0.001$) with testis weight, with the variation in AGD accounting for 23%, 59% and 32% of the variation in testis weight at e21.5, postnatal day 25 and adulthood, respectively (Fig. 3). Of these three ages, postnatal day 25 is the most informative, as final Sertoli cell number has been determined by this age but other (confounding) factors that can affect adult testis weight, such as cryptorchidism which is common in DBP-exposed rats, have no significant impact at this age (Hutchison et al. 2008). Indeed, at postnatal day 25, a significant correlation ($P = 0.002$) between AGD and testis weight was even found just in the control group, despite the much narrower range of values (Fig. 3). Sertoli cell number was determined at postnatal day 25 in a subset of control and DBP-exposed males and was shown to correlate highly significantly ($P < 0.001$) with AGD, with variation in the latter explaining 42% of the variation in (final) Sertoli cell number (Fig. 4).

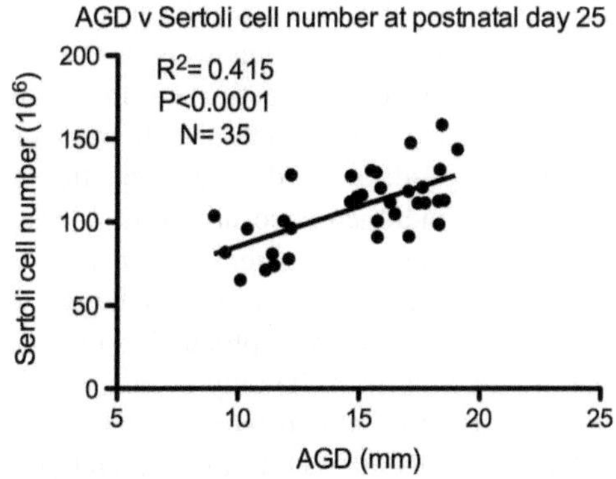

Fig. 4 Relationship between anogenital distance (AGD) and final Sertoli cell number at postnatal day 25 in male offspring from mothers subjected to one of the treatment modalities listed in the first four rows of Table 1. Correlations were determined using linear regression analysis. Sertoli cell number was determined as detailed by Auharek et al. (2010)

Discussion and Conclusions

The present studies provide proof of concept in relation to the time window-specific programming of AGD by androgens in the rat, and they demonstrate in large experimental datasets that AGD is strongly and positively correlated to testis size, both during growth and in adulthood. Although we have shown both of these concepts in earlier studies (Welsh et al. 2008; Scott et al. 2008; Drake et al. 2009), the present data go some way beyond these earlier studies because they are more detailed and involve a much wider spectrum of experimental treatments. Earlier studies that sought to experimentally disrupt steroidogenic function by the fetal rat testis had exclusively used DBP. As this compound can directly target Sertoli cells (Foster 2006), the concern was that any relationship between AGD and testis weight in such studies could have been due to the fortuitous association of two direct, independent effects of the DBP, one on testosterone production (which then impacted AGD and perhaps also Sertoli cell number/proliferation) and one on Sertoli cell number/proliferation. We therefore used a wider range of experimental chemical treatments based on reports in the literature that showed inhibition of steroidogenesis by the fetal testis by prochloraz (Blystone et al. 2007) and linuron (Wilson et al. 2009). By using these compounds as well as DBP, and including combinations of these chemicals, it was considered that any observed relationship between AGD and testis weight would have to reflect a causative, rather than a coincidental, association.

Using the treatments described above (detailed in Table 1), our results show an unequivocally strong relationship between AGD and testis weight in datasets comprising >150 animals (a mixture of control and compound-treated animals). This relationship was equally evident in late fetal life, during puberty and in adulthood. At postnatal day 25, when the strongest AGD-testis weight relationship was found, a significant correlation was even found in the control group, implying that even normal variation in testis weight may result from normal variation in androgen action and programming during the MPW. Furthermore, for a subset of control/chemical-treated animals, there was an equally strong correlation of AGD with Sertoli cell number at this age. This finding contrasts with our earlier findings (Scott et al. 2008), but it was based on a much smaller sample size. This is a logical finding in that it is well established that variation in Sertoli cell number is the primary determinant of adult testis size (Sharpe et al. 2003). In the treatments listed in Table 1, we did not directly measure intratesticular testosterone levels during the MPW; instead we derived these data from the measurement of AGD. In view of the absolute dependence of data interpretation on the reliability of AGD as a specific read-out of androgen exposure in the MPW, we also undertook further validation studies to confirm our earlier findings, as most of these had relied on antagonism of peripheral androgen action with flutamide (Welsh et al. 2008, 2010) rather than disruption of testosterone production by the fetal testis. Our findings show clearly that a reduction in AGD only occurs when testosterone production is suppressed within the MPW. It can therefore be concluded that androgen action within the MPW programs ultimate testis size, and therefore sperm production, via unknown mechanisms. The present data point to this programming effect being mediated via an effect on Sertoli cell number/proliferation. How this mediation operates is a mystery, especially as Sertoli cell proliferation continues well beyond the confines of the MPW and may be driven by androgens outside of the MPW (Scott et al 2007, 2008; Auharek et al 2010). Determination of this mechanism is a priority in view of the human health implications of these findings.

At the present time, average sperm counts in young men across Northern Europe are remarkably low, and an alarmingly high percentage (~20%) have categorically abnormal sperm counts (<20 million/mL ejaculate) that are likely to impact adversely on fertility (Jørgensen et al. 2006; Skakkeback et al. 2007). The best data suggest that, 60 years ago, average sperm counts were much higher, with a much lower percentage of men having abnormally low sperm counts (Carlsen et al 1992; Swan et al. 2000). Based on the TDS hypothesis and the established strong relationship between Sertoli cell number and testis size and sperm counts in men (Sharpe et al. 2003), it is reasonable to surmise that the changes in sperm counts could reflect altered androgen action within the MPW. There are currently no available data on this hypothesis (although studies are in the pipeline), but in humans, as in rats, AGD has already been shown to correlate with penis length and the occurrence of hypospadias and cryptorchidism (Swan et al. 2005; Hsieh et al. 2008; Thankamony et al. 2009). If, as predicted, AGD is found to correlate with testis size/sperm counts in humans, and if in particular reduced AGD is associated with low/reduced sperm counts, these findings will have considerable implications. They will mean that, just as predicted in the TDS hypothesis, declining sperm counts in humans has its origins in fetal life and is probably preventable. The findings also mean that adult sperm counts will to an extent be predictable from measurement of AGD at birth or prior to puberty.

Acknowledgments This work was supported in part by the UK Medical Research Council (WBS U.1276.00.001.00038.02) and by the European Union Framework 7 programme (DEER; FP7-ENV-2007-1-212844).

References

Andersson AM, Jørgensen N, Main KM, Toppari J, Rajpert-De Meyts E, Leffers H, Juul A, Jensen TK, Skakkebaek NE (2008) Adverse trends in male reproductive health: we may have reached a crucial 'tipping point'. Int J Androl 31:74–80

Auharek SA, de Franca LR, McKinnell C, Jobling MS, Scott HM, Sharpe RM (2010) Prenatal plus postnatal exposure to di(n-butyl) phthalate and/or flutamide markedly reduces final Sertoli cell number in the rat. Endocrinology 151:2868–2875

Blystone CR, Lambright CS, Howdeshell KL, Furr J, Sternberg RM, Butterworth BC, Durhan EJ, Makynen EA, Ankley GT, Wilson VS, LeBlanc GA, Gray LE Jr (2007) Sensitivity of fetal rat testicular steroidogenesis to maternal prochloraz exposure and the underlying mechanism of inhibition. Toxicol Sci 97:512–519

Carlsen E, Giwercman A, Keiding N, Skakkebaek NE (1992) Evidence for decreasing quality of semen during past 50 years. BMJ 305:609–613

Drake AJ, Van den Driesche S, Scott HM, Hutchison G, Seckl JR, Sharpe RM (2009) Glucocorticoids amplify dibutyl phthalate-induced disruption of fetal testosterone production and male reproductive development. Endocrinology 150:5055–5064

Fisher JS, Macpherson S, Marchetti N, Sharpe RM (2003) Human 'testicular dysgenesis syndrome': a possible model based on in utero exposure of the rat to dibutyl phthalate. Hum Reprod 18:1383–1394

Foster PM (2006) Disruption of reproductive development in male rat offspring following in utero exposure to phthalate esters. Int J Androl 29:140–147

Gray LE Jr, Wilson VS, Stoker T, Lambright C, Furr J, Noriega N, Howdeshell K, Ankley GT, Guillette L (2006) Adverse effects of environmental antiandrogens and androgens on reproductive development in mammals. Int J Androl 29:96–104

Hsieh MH, Breyer BN, Eisenberg ML, Baskin LS (2008) Associations among hypospadias, cryptorchidism, anogential distance, and endocrine disruption. Curr Urol Rep 9:137–142

Hutchison G, Scott HM, Walker M, McKinnell C, Mahood IK, Ferrara D, Sharpe RM (2008) Sertoli cell development and function in an animal model of testicular dysgenesis syndrome. Biol Reprod 78:352–360

Jørgensen N, Asklund C, Carlsen E, Skakkebaek NE (2006) Coordinated European investigations of semen quality: results from studies of Scandinavian young men is a matter of concern. Int J Androl 29:54–61

MacLeod DJ, Sharpe RM, Welsh M, Fisken M, Scott HM, Hutchison GR, Drake AJ, van den Driesche S (2010) Effect of disruption of androgen production or action in the masculinisation programming window on the development of male reproductive organs. Int J Androl 33:279–287

Scott HM, Hutchison GR, Mahood IK, Hallmark N, Welsh M, de Gendt K, Verhoeven G, O'Shaughnessy PJ, Sharpe RM (2007) Role of androgens in fetal testis development and dysgenesis. Endocrinology 148:2027–2036

Scott HM, Hutchison GR, Jobling MS, McKinnell C, Drake AJ, Sharpe RM (2008) Relationship between androgen action in the 'male programming window', fetal Sertoli cell number and adult testis size in the rat. Endocrinology 149:5280–5287

Scott HM, Mason JI, Sharpe RM (2009) Steroidogenesis in the fetal testis and its susceptibility to disruption by exogenous compounds. Endocr Rev 30:883–925

Sharpe RM (1994) Regulation of spermatogenesis. In: Knobil E, Neill JD (eds) The physiology of reproduction, 2nd edn. Raven, New York, pp 1363–1434

Sharpe RM, Skakkebaek NE (2008) Testicular dysgenesis syndrome: mechanistic insights and potential new downstream effects. Fertil Steril 89(Suppl 1):e33–e38

Sharpe RM, McKinnell C, Kivlin C, Fisher JS (2003) Proliferation and functional maturation of sertoli cells, and their relevance to disorders of testis function in adulthood. Reproduction 125:769–784

Skakkebaek NE, Rajpert-De Meyts E, Main KM (2001) Testicular dysgenesis syndrome: an increasingly common developmental disorder with environmental aspects. Hum Reprod 16:972–978

Skakkebaek NE, Rajpert-De Meyts E, Jørgensen N, Main KM, Leffers H, Andersson AM, Juul A, Jensen TK, Toppari J (2007) Testicular cancer trends as 'whistle blowers' of testicular developmental problems in populations. Int J Androl 30:198–204

Swan SH, Elkin EP, Fenster L (2000) The question of declining sperm density revisited: an analysis of 101 studies published 1934–1996. Environ Health Perspect 108:961–966

Swan SH, Main KM, Liu F, Stewart SL, Kruse RL, Calafat AM, Mao CS, Redmon B, Ternand CL, Sullivan S, Teague JL, the Study for Future Families Research Team (2005) Decrease in anogenital distance among male infants with prenatal phthalate exposure. Environ Health Perspect 113:1056–1061

Thankamony A, Ong KK, Dunger DB, Acerini CL, Hughes IA (2009) Anogenital distance from birth to 2 years: a population study. Environ Health Perspect 117:1786–1790

van den Driesche S, Scott HM, MacLeod DJ, Fisken M, Walker M, Sharpe RM (2011) Relative importance of prenatal and postnatal androgen action in determining growth of the penis and anogenital distance (AGD) in the rat before, during and after puberty. Int J Androl doi: 10.1111/j.1365-2605.2011.01175.x

Welsh M, Saunders PTK, Fisken M, Scott HM, Hutchison GR, Smith LB, Sharpe RM (2008) Identification in rats of a programming window for reproductive tract masculinization, disruption of which leads to hypospadias and cryptorchidism. J Clin Invest 118:1479–1490

Welsh M, MacLeod D, Walker M, Saunders PTK, Sharpe RM (2010) Critical androgen-sensitive periods of development and growth of the rat penis and clitoris. Int J Androl 33:e144–e152

Wilson VS, Lambright CR, Furr JR, Howdeshell KL, Gray LE Jr (2009) The herbicide linuron reduces testosterone production from the fetal rat testis during both in utero and in vitro exposures. Toxicol Lett 186:73–77

Index

J.-P. Bourguignon et al. (eds.), *Multi-System Endocrine Disruption*, Research and Perspectives in Endocrine Interactions, DOI 10.1007/978-3-642-22775-2, © Springer-Verlag Berlin Heidelberg 2011

MIX
Papier aus verantwortungsvollen Quellen
Paper from responsible sources
FSC® C105338

If you have any concerns about our products,
you can contact us on
ProductSafety@springernature.com

In case Publisher is established outside the EU,
the EU authorized representative is:
Springer Nature Customer Service Center GmbH
Europaplatz 3, 69115 Heidelberg, Germany

Printed by Libri Plureos GmbH
in Hamburg, Germany